MICROWAVE TECHNIQUES

PHILIPS TECHNICAL LIBRARY

MICROWAVE TECHNIQUES

H. Mooijweer

MACMILLAN

SBN 333 12030 2

First published 1971 *by*
THE MACMILLAN PRESS LTD
London and Basingstoke
*Associated companies in New York Toronto Melbourne
Dublin Johannesburg and Madras*

PHILIPS

Trademarks of N.V. Philips' Gloeilampenfabrieken

Printed in Great Britain at The Whitefriars Press, London and Tonbridge

FOREWORD

After an introduction concerning the necessity of using Maxwell's equations when dealing with phenomena in the microwave frequency range, the theoretical and experimental basis of electromagnetism is briefly summarized in the first part of this book (field strength, dielectric displacement, electric flux, Biot and Savart's law, magnetic flux, Faraday's law of induction). Maxwell's well-known laws of electromagnetism are then derived from the foregoing, in the integral form in which he first proposed them. The concept of the propagation of electromagnetic energy in the form of waves is illustrated with reference to these laws in their integral form. They are then rewritten in differential form, and the general boundary conditions for the electromagnetic field at the surface of conductors and the interface between dielectrics are discussed.

Consideration of Maxwell's equations for an unbounded dielectric (for the sake of simplicity, the dielectric is assumed lossless, and the electric field is assumed to have only one non-zero component, which is a function of time and of one position coordinate only) leads to the wave equation, with travelling plane waves as solutions; the propagation velocity, dispersion and polarization of these waves are discussed, together with the Poynting vector as a measure of the energy content of the waves.

The propagation of electromagnetic waves in a conducting medium is then discussed, and it is found that the basic wave type here is the linearly polarized plane wave. The complex notation for electromagnetic waves is introduced at this point; the theory of complex numbers is briefly reviewed in an appendix.

The solution of the wave equation in this case leads to damped travelling waves (with attenuation constant and phase constant). We then consider the transmission and reflection of plane polarized electromagnetic waves, normally incident on the interface between air and a perfect conductor, air and an ideal dielectric, and air and a lossy conductor. The concepts of standing waves (nodes and anti-nodes), the standing-wave ratio and the reflection coefficient are introduced at this stage.

v

The above-mentioned basic theory takes up about one third of the book: the rest is concerned with waveguides, being divided more or less equally between parallel-wire transmission lines and rectangular waveguides.

After a brief introduction to the concept of waveguides in general, we start by considering parallel-wire transmission lines (coaxial line, Lecher line, etc.). The propagation of the waves along these lines (by the TEM mode) is discussed with reference to the 'telegraphers' equations', and we are introduced here to voltage and current waves, the characteristic impedance, the propagation constant, phase velocity, group velocity, travelling and standing waves, reflection coefficient in waveguides, line segments of finite length (not terminated by a matched load), and input impedance. Various applications of lengths of parallel-wire transmission line (open or short-circuited) are then discussed: as self-inductances or capacitances in the UHF range, as wave-meter, as impedance transformer, etc. This section of the book is closed by a discussion of a number of components used in these lines (directional couplers, short-circuit plungers, attenuators, matched loads, power dividers, etc.).

The Smith chart, a useful graphical aid to the calculation of both parallel-wire transmission lines and rectangular waveguides, is discussed in an appendix.

The third part of the book deals with rectangular waveguides. We start with a theoretical treatment of the propagation of electromagnetic waves in a rectangular waveguide (solution of Maxwell's equations with the appropriate boundary conditions; solution of the wave equations by separation of the variables; the TE and TM modes; propagation constant; cut-off frequency; principal mode; phase and group velocity in waveguides; waveguide wavelength). The concept of impedance in waveguides is developed by analogy with that for parallel-wire transmission lines; we find that no absolute value of the impedance can be defined, and normalized impedances are introduced to deal with this.

Wave propagation in round waveguides and the higher modes in coaxial lines are mentioned in passing. As with parallel-wire transmission lines, we then deal with the various waveguide components, after an introduction to the information in waveguide handbooks about the more general types of discontinuities such as inductive and capacitive diaphragms. The other components dealt with include impedance transformers (sliding screw, E-H tuners, quarter-wave transformers, tapers, etc.), short-circuiting plungers, bends, transitions from waveguides to coaxial lines, attenuators, directional couplers, matched loads, T-junctions and filters.

Finally, some attention is devoted to cavity resonators. The generation and amplification of microwaves (klystron, magnetron, travelling-wave tube, etc.) and their radiation (antennae) are not discussed in this book.

Modern components based on magnetic and semiconducting materials are mentioned in an appendix.

The main aim has been to give a general picture of microwave techniques, while avoiding the reference to vector analysis which makes most of the literature on this subject so difficult to some readers.

H. MOOIJWEER.

CONTENTS

Foreword   v

1. Introduction   1

Electromagnetic fields in unbounded media
2. Basic experiments   7
3. Maxwell's equations   20
   3.1 Maxwell's equations in integral form   20
   3.2 Maxwell's equations in differential form   22
4. Boundary conditions for electromagnetic fields   27
   4.1 Interfaces between dielectrics   27
   4.2 Interface between dielectrics and ideal conductors   31
5. Electromagnetic waves in unbounded dielectrics   33
   5.1 The wave equation for plane waves   33
   5.2 Polarization of waves   40
   5.3 The Poynting vector   41
6. Electromagnetic waves in conducting media   45
   6.1 Complex notation   45
   6.2 Propagation of electromagnetic waves in unbounded
       conducting media   48
7. Reflection and transmission of electromagnetic waves at
   interfaces   52
   7.1 Electromagnetic waves from free space incident on ideal
       conductors   52
   7.2 Electromagnetic waves from free space incident on perfect
       dielectrics   56
   7.3 Electromagnetic waves from free space incident on media
       of finite conductivity   60

Wave bearing systems
8. General considerations   63

Parallel-wire transmission lines
9. Derivation of the telegraphers' equations ... 73
   9.1 The propagation constant ... 78
   9.2 Phase velocity and group velocity ... 79
10. General properties of parallel-wire transmission lines ... 82
   10.1 Infinitely long transmission lines; characteristic impedance ... 82
   10.2 Calculation of line constants L and C ... 85
   10.3 Transmission lines of finite length; standing waves and reflection ... 88
   10.4 Input impedance ... 93
11. Applications of transmission lines ... 96
   11.1 Transmission lines as UHF self-inductances or capacitances ... 96
   11.2 Transmission lines as resonant circuits ... 99
   11.3 Transmission lines as wavemeters ... 105
   11.4 Impedance measurements with transmission lines ... 107
   11.5 Transmission lines as impedance transformers ... 109
   11.6 Discontinuities ... 115
12. Components in parallel-wire transmission lines ... 119
   12.1 Supports ... 119
   12.2 Short-circuit plungers ... 121
   12.3 Matched loads ... 123
   12.4 Attenuators ... 125
   12.5 Detectors ... 126
   12.6 Directional couplers ... 128
   12.7 Filters ... 130
   12.8 Power dividers ... 131
   12.9 Measurement techniques ... 136

Waveguides
13. Wave propagation in rectangular waveguides ... 143
   13.1 Introduction ... 143
   13.2 TE or H waves ... 146
   13.3 TM or E waves ... 151
14. Properties of rectangular waveguides ... 156
   14.1 Propagation constant and cut-off frequency ... 156
   14.2 Phase velocity and group velocity; guide wavelength ... 160
   14.3 Standing waves and reflection ... 162
   14.4 Impedance in waveguides ... 166
   14.5 Attenuation in waveguides ... 173
15. Wave propagation in circular waveguides ... 175
16. Higher modes in coaxial lines ... 177

17. Waveguide components   179
    17.1 Discontinuities   179
    17.2 Impedance transformers   188
    17.3 Short-circuit plungers   194
    17.4 Bends   194
    17.5 Transitions from waveguide to coaxial line   197
    17.6 Detection   199
    17.7 Standing-wave detectors   200
    17.8 Phase shifters   204
    17.9 Attenuators and matched loads   205
    17.10 Directional couplers   206
    17.11 T-junctions   209
    17.12 Power dividers   213
    17.13 Filters   215
    17.14 Flanges   221

Cavity resonators
18. Introduction to the cavity resonator   225
19. Simple cavity resonators   228
    19.1 Introduction   228
    19.2 Rectangular parallelepiped   231
    19.3 Circular cylinders   236
    19.4 Coaxial cavity resonators   243
20. Equivalent circuits and coupling coefficients of cavity resonators 244

Appendices
    Appendix 1 Complex numbers   253
    Appendix 2 The Smith chart   266
    Appendix 3 Solid-state microwave components   276
    Appendix 4 Frequency ranges and waveguide dimensions   286
    Bibliography for further reading   289
    Index   291

1 INTRODUCTION

Before we begin our study of Maxwell's equations and their application to various high-frequency problems, we should consider why such complicated procedures are necessary in the microwave frequency range. At low frequencies, even the most complicated circuit can be described in terms of simple concepts such as self-inductance, capacitance, resistance and the like, and this description often enables all the problems met with in practice to be solved satisfactorily.

The main reason lies in the dimensions of the circuits and their components compared with the wavelength of the current flowing through them. This is illustrated in the table below, which is based on the well-known relationship between the wavelength and frequency of electrical phenomena

$$\lambda = \frac{c}{f} \tag{1.1}$$

where c = velocity of light in air $(3 \times 10^8$ m/s), λ = the wavelength in metres, and f = the frequency in cycles per second (Hertz).

	f (Hz)	λ (m)	
AC mains	50	6×10^6	⎫ Conventional circuits;
Radio	10^6	300	⎭ lumped elements
TV, FM	10^8–10^9	3–0·3	Transmission lines (coaxial cables, etc.); distributed elements
Radar	10^9–10^{10}	0·3–0·03	Waveguides

At low frequencies (AC mains and, to some extent, radio waves), the wavelength is nearly always much larger than the dimensions of the circuit. In this range, we can work with L, C, R networks in the conventional way, considering self-inductances, capacitances and resistances as 'lumped elements' since at the wavelengths we are dealing with, we

1

can really conceive of a self-inductance concentrated at one particular point, a capacitance concentrated at another, etc. The situation starts to change at high radio frequencies, and is certainly different in the FM and television range. It will be remembered that at these frequencies a coil actually has a certain capacitance per unit length because of the field between neighbouring turns, as well as a self-inductance; this capacitance is certainly not negligible at high frequencies (Fig. 1.1).

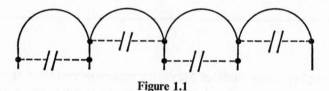

Figure 1.1

At slightly higher frequencies (i.e. lower wavelengths, of the order of a few decimetres), the largest dimension of the conductors (i.e. their length) will be of the same order of magnitude as the wavelength, and may even exceed the latter. We are now in the range of the coaxial cable and similar parallel-wire transmission lines. However, the transverse dimensions of the transmission line are generally still small compared with the wavelength. This means that we can still speak of the voltage between the inner and the outer conductor of a coaxial cable in a given cross-section, but we can no longer consider self-inductance and capacitance as being separated in space. If we so wish, we can regard the self-inductance and capacitance as being distributed along the transmission line. The line thus has a certain self-inductance and a certain capacitance per unit length, and can be represented schematically as a linked chain of self-inductances and capacitances (Fig. 1.2)—as long as it is lossless. In this range, we are thus dealing with 'distributed elements'.

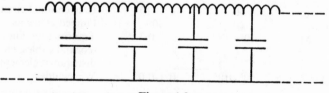

Figure 1.2

The range where coaxial cables are used can be considered as being transitional to that in which waveguides are often used as conductors. Here, even the transverse dimensions of the conductor are of the same order of magnitude as the wavelength, and concepts such as self-inductance and capacitance have lost their original meaning. The result of this is that we can no longer use these concepts for calculations as we could at low frequencies.

Both with a parallel-wire transmission line and with a waveguide, the effects of a frequency voltage applied at a given point of the circuit are not immediately felt throughout the circuit. A finite time is required for the changes in the voltage (the electric field) to become evident at different points down the line; the disturbance is propagated down the line at a finite velocity, in the form of a wave. In principle, this is also true of a low-frequency circuit whose dimensions are small compared with the wavelength, but the time lag is not noticed in practice. The current behaves in a similar way: in an LCR network with lumped elements, the current in a closed circuit has the same amplitude and phase everywhere—at least, the error made by assuming this is generally negligible. The effects are quite different at very high frequencies, where the energy travels along the conductors at a finite velocity. There will now be a phase shift, or time lag, between the currents (or voltages) at different points in the circuit. It will be obvious from the above description that there is no clear boundary line between the lumped-element range and the distributed-element range.

The consequence of these facts will be clear. In a system where very high frequencies are involved and where we can no longer speak of clearly-defined currents, voltages, self-inductances, etc., we must solve the equations which are generally applicable to such systems (Maxwell's equations) with reference to the boundary conditions of the system in question.

It would be an impossible task to re-solve Maxwell's equations afresh each time a small change was made in the circuit. Fortunately, microwave circuits (we are not speaking here about the generation and radiation of microwaves) do not vary as much as the circuits for lower frequencies. They are mainly built up of various kinds of cavity resonators and transmission lines. Since the properties of such resonators and lines can be determined in fairly general terms, it is not necessary to repeat the time-consuming procedure of solving Maxwell's equations each time, say, the dimensions of a resonator are changed. Difficulties do arise when wave-guides of different (transverse) dimensions have to be connected, or when other discontinuities are present in the transmission line so that these lines are no longer uniform. However, if the configuration is not too complicated, general rules governing these discontinuities can be derived and laid down in tabular or graphical form for use in special cases.

It will be clear from the above that it is indeed worth while to become familiar with Maxwell's equations and their manipulation in order to be able to solve certain microwave problems.

ELECTROMAGNETIC FIELDS
IN UNBOUNDED MEDIA

2 BASIC EXPERIMENTS

Before introducing Maxwell's equations, we shall review the basic equations and experiments relating to electric and magnetic fields. It is assumed that the reader will be acquainted with such concepts as field strength, lines of force, potential, Coulomb's law, and other similar concepts.

Experiment 1

Every electrically charged body is surrounded by a field of force (Fig. 2.1). This field may be mapped with a minute sphere carrying a small

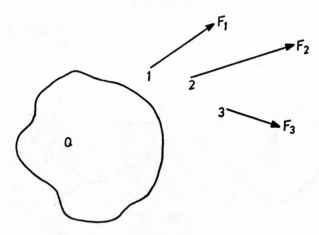

Figure 2.1

electric charge which is moved through the region in question. The force on this sphere at point 1 is found to be F_1, that at point 2 is F_2, and so on. The force is found to be proportional to the charge on the 'test sphere'. This interplay of forces or change in state produced by the presence of a charged body, is called an electric field.

By definition, the electric field strength at a given point is equal in magnitude and direction to the force on a test sphere of positive unit charge placed at that point

$$F = QE \tag{2.1}$$

where F is the force in newtons, E is the field strength in volts per metre, and Q is the charge (on the test sphere) in coulombs.

The direction of the field strength is the direction of the force.

Experiment 2

E is an electrostatic field produced by an arbitrary body $\mathscr{L}$ of charge Q. If a test sphere of unit charge is moved round a closed path in this field, it is found that the total work done is zero.

$$\lim_{\Delta l \to 0} \sum F_l . \Delta l = \oint F_l \, dl = 0 \tag{2.2}$$

F_l is the component of the force F exerted by the field on the test sphere in the direction of the segment Δl of the closed curve in question; $\oint$ indicates integration round a closed curve. Since, as we have seen in experiment 1, the electrical field is always proportional to the force, we may also write for an electrostatic field

$$\oint E_l \, dl = 0 \tag{2.3}$$

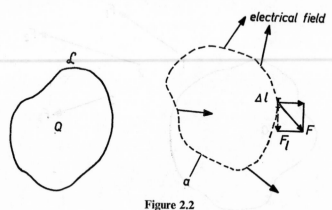

Figure 2.2

Experiment 3

Consider an arbitrary, closed, imaginary surface S in a space subject to an electric field E. We now measure the electric field strength at all points on this surface with the aid of a test sphere. In fact, we imagine the surface S divided up into very small parts ΔS, and determine the field strength on each ΔS (Fig. 2.3). We shall denote the component of E perpendicular to ΔS by E_n (Fig. 2.4). E_n is taken as positive when it is directed outward, and negative when it is directed inward. We now

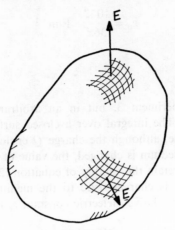

Figure 2.3

calculate the product $E_n.\Delta S$ for each surface element ΔS, summed over the whole surface $(\sum E_n.\Delta S)$ and take the limit as $\Delta S \to 0$

$$\lim_{\Delta S \to 0} \sum E_n \Delta S = \int_S E_n \, dS \qquad (2.4)$$

Now the magnitude of this integral is found to be proportional to the total electric charge Q enclosed by the closed surface S.

$$\varepsilon_0 \int_S E_n \, dS = Q \quad (in \; vacuo) \qquad (2.5)$$

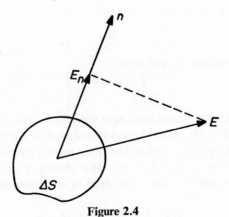

Figure 2.4

If this integral over the closed surface is zero, it follows that there is no charge within the surface.

The magnitude of the proportionality constant ε_0 depends on the units used. In the rationalized MKSA system, which we shall be using in this book, we have

$$\varepsilon_0 = \frac{10^{-9}}{36\pi} \quad F\ m^{-1} \tag{2.6}$$

Experiment 4

We now repeat experiment 3, but in an arbitrary medium (e.g. oil) instead of *in vacuo*. The integral over a closed surface is now found to have a different value, although the charge Q enclosed by the surface S is the same. If the medium is changed, the value of the integral will also change. In order to retain the validity of equation (2.5) we have to introduce a factor which is characteristic to the medium in question. This constant is called the relative dielectric constant ε_r for that medium. We now write

$$\varepsilon_0 \varepsilon_r \int_{\substack{\text{closed} \\ \text{surface}}} E_n\ dS = Q \tag{2.7}$$

The magnitude of ε_r depends on the medium, the temperature and other parameters. In air, $\varepsilon_r = 1\cdot0006 \approx 1$, so that air and vacuum do not differ appreciably in this respect. Pure distilled water has $\varepsilon_r \approx 80$, and teflon has $\varepsilon_r \approx 2\cdot25$.

If various different materials are cut by the surface S, it is better to write

$$\int_{\substack{\text{closed} \\ \text{surface}}} \varepsilon_0 \varepsilon_r E_n\ dS = Q \tag{2.8}$$

since ε_r also varies during the integration.

If we call

$$\varepsilon = \varepsilon_0 \varepsilon_r \tag{2.9}$$

the dielectric constant, we can write

$$\int_{\substack{\text{closed} \\ \text{surface}}} \varepsilon E_n\ dS = Q \tag{2.10}$$

We can now formally introduce a new variable D, the dielectric displacement. (Historically, this variable was introduced in a different way.) This quantity has a magnitude and a direction; in other words, it is a vector. The dielectric displacement has the same direction as the electric field strength and is equal to ε times the latter in any medium. We thus have, by definition,

$$D = \varepsilon E = \varepsilon_0 \varepsilon_r E \quad C\ m^{-2} \tag{2.11}$$

Equation (2.10) can therefore be rewritten as

$$\int\limits_{\substack{\text{closed}\\\text{surface}}} D_n\, dS = Q \qquad (2.12)$$

This may be expressed in words as follows: the integral over an arbitrary closed surface of the electric flux ψ is equal to the total charge enclosed by the surface.

This statement introduces a new phrase: the product $D_n\, dS$ is called the electric flux through the surface element dS (Fig. 2.5). Here again,

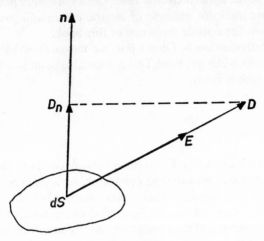

Figure 2.5

D_n is positive when it points outwards and negative when it points inwards.

$$\psi = \int D_n\, dS \qquad (2.13)$$

Equation (2.13) represents the electric flux through a surface S (which need not be closed in this equation).

Note: The basic experiments presented above form a sufficient basis for the whole of electrostatics, but it is equally possible to choose another line of argument and other basic experiments (e.g. based on Coulomb's law). The procedure followed historically differed from the above in many respects, but is not always the best from an educational point of view.

Experiment 5

If we apply an electric field to a conductor, charge will flow through the conductor. This flow of charge is called an electric current, and is defined as the amount of charge flowing through a cross-section of the conductor in unit time

$$I = \frac{dQ}{dt} \qquad (2.14)$$

The fifth experiment consists of the work carried out by Ohm leading to Ohm's law:

$$J = \gamma E \tag{2.15}$$

where E is the applied electric field in volts per metre in the conductor, J is the current density in amps per square metre at a point in the conductor and γ is the conductivity of the conductor in siemens per metre (a proportionality constant which depends on the material). Apart from the magnitude given in equation (2.15), the current density also has a direction, namely that of the applied electric field. Ohm's law does not apply under all conditions: think for example of electric currents in gases. However, such conditions are outside the scope of this book.

With the introduction of Ohm's law, we moved from electrostatics to situations in which charges flow. This is a suitable point to introduce some aspects of magnetic fields.

Experiment 6

If a conductor through which current is flowing comes in the neighbourhood of another current-carrying conductor, or a magnet, the two will exert forces on one another. These forces cease when the current stops flowing. Various experiments can be carried out in which the magnitude of the current, the size of the conductor, the distance from the conductor and the angle of the conductor are varied. The results of these experiments can be summarized in Biot and Savart's law, which states that a magnetic field of a certain magnitude and direction exists round a current-carrying conductor (further details are given below; see Fig. 2.6).

This law is probably best known in its differential form. If P is a point at distance l from the segment ΔS of the conductor in question, and if ΔS makes an angle φ with the line joining its centre to P and carries a current i, then according to Biot and Savart's law the magnetic field strength ΔH_P at the point P due to the segment ΔS of the conductor is given by

$$\Delta H_P = \frac{i\Delta S}{4\pi l^2} \sin \varphi \quad \text{A m}^{-1} \tag{2.16}$$

and is perpendicular to the plane formed by ΔS and l.

The magnetic field strength at P due to the whole conductor can be found by integrating equation (2.16) over the length of the conductor. This gives the magnitude of the magnetic field strength. The direction of H is given by the 'right-hand' or corkscrew rule; it is the direction in which the point of a corkscrew with a right-hand thread would move if the handle were turned through the smallest possible angle from the

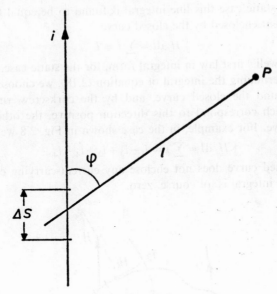

Figure 2.6

positive direction of the current to the point P. In the example of Fig. 2.6, this direction is perpendicular to the plane of the paper and into it.

Experiment 7

Biot and Savart's law is not used in the above form in microwave theory. As in experiment 2 for the electrostatic field, we can now for the magnetostatic field integrate H_l round a closed curve (Fig. 2.7). This gives

$$\lim_{\Delta l \to 0} \sum H_l . \Delta l = \oint H_l \, dl \qquad (2.17)$$

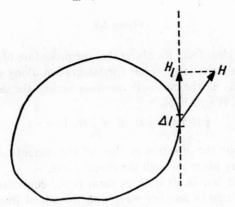

Figure 2.7

Now in the static case this line integral is found to be equal to the total electric current enclosed by the closed curve

$$\oint H_l \, dl = \sum i = I \tag{2.18}$$

This is Maxwell's first law in integral form, for the static case.

When determining the integral of equation (2.18), we choose a positive direction round the closed curve, and by the corkscrew rule call the currents which correspond to this direction positive, the other currents being negative. For example, in the case shown in Fig. 2.8 we may write

$$\oint H_l \, dl = \sum i = i_i - i_2 + i_3 + i_4 - i_4 \tag{2.19}$$

If the closed curve does not enclose any current-carrying conductors, then the line integral is, of course, zero.

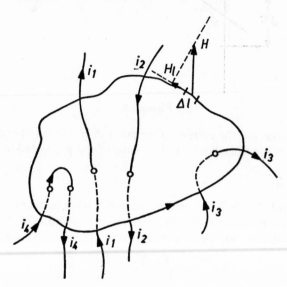

Figure 2.8

It will also be clear from the above that the procedure of equation (2.18) will give precisely the same result for integration along different closed curves, as long as all the curves in question enclose the same conductors in the same way (Fig. 2.9)

$$\oint_{c1} H_l \, dl = \oint_{c2} H_l \, dl = \oint_{c3} H_l \, dl = -i \tag{2.20}$$

In order to define the direction of flow of the currents clearly, we may draw an arbitrary plane through the closed curve.

Maxwell's first law in its integral form is the equivalent of Biot and Savart's law written in another way, with the added freedom of choice of the integration path as indicated above.

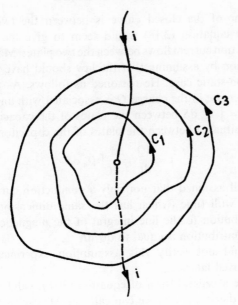

Figure 2.9

Experiment 8

We shall now show that equation (2.18) really is restricted to static fields. Let us consider a capacitor being charged by a current $i = \mathrm{d}Q/\mathrm{d}t$, where Q is the charge on the capacitor (Fig. 2.10). If we now apply equation (2.18) to the curve c, where the plane containing c cuts the lead wires to the capacitor in A or B, say, then we find $\oint H_t \, \mathrm{d}l = i$. If, on the other

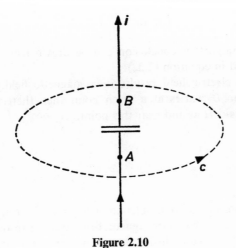

Figure 2.10

hand, the plane of the closed curve is between the two plates of the capacitor, then equation (2.18) would seem to give the result of zero, since no conduction current flows between the two plates. Maxwell explained this contradiction by assuming that the law should have a more general form in the non-static case. He reasoned as follows. As the capacitor is charged, the increase in the charge Q is associated with an increase in the electric flux $\psi = \int D_n \, dS$ between the plates of the capacitor.

For a plane situated between the plates of the capacitor, we may write

$$i = \frac{dQ}{dt} = \frac{d}{dt} \int D_n \, dS \tag{2.21}$$

Now Maxwell assumed that not only a conduction current but also a flux ψ varying with time (which has the same dimension as a current) makes a contribution to the line integral of the magnetic field strength, and that this contribution is equal to $d\psi/dt$.

Maxwell could not verify this assumption experimentally, but its validity was proved later.

We now have a general form of equation (2.18), valid in the dynamic case as well as the static, and we can call this Maxwell's first law

$$\oint H_l \, dl = i + \frac{d}{dt} \int_S D_n \, dS \tag{2.22}$$

where i is the total conduction current enclosed by the closed curve round which the magnetic field strength is integrated, and the surface S over which the dielectric displacement is to be integrated is one with this closed curve as its boundary.

In free space (where there are no conduction currents) equation (2.22) becomes

$$\oint H_l \, dl = \frac{d}{dt} \int_S D_n \, dS \tag{2.23}$$

The following important conclusion can be drawn from Maxwell's first law as expressed in equation (2.22):

A varying electric field results in a magnetic field, such that when E (or D) varies at a given point then there must be a magnetic field H at and near this point.

Experiment 9

Let us now consider a long solenoid carrying a current i. It follows from equation (2.18) that the magnetic field strength in the coil does not change if the air or vacuum inside the coil is replaced by another material.

However, different materials do cause some difference, namely in the magnetic flux Φ. This flux is measured as an induced e.m.f. in a test coil connected to a ballistic galvanometer, when the current in the solenoid is commutated (Fig. 2.11).

The magnetic flux when there is a medium in the coil is found to be a factor μ_r greater than *in vacuo*. This factor μ_r is called the relative permeability of the medium.

For most substances, $\mu_r \approx 1$. For diamagnetic materials like copper, μ_r is slightly less than 1, and for paramagnetic materials like aluminium it is slightly greater than 1. However, for ferromagnetic materials like iron, μ_r is much greater than 1.

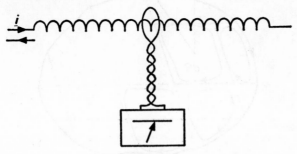

Figure 2.11

We may now introduce the magnetic flux density or magnetic induction B as a measure of the magnetic flux. This quantity has the same direction as the magnetic field strength; its magnitude is defined as follows:

$$B = \mu H = \mu_0 \mu_r H \quad \text{Wb m}^{-2} \qquad (2.24)$$

where $\mu_0 = 4\pi . 10^{-7}$ H/m in the rationalized MKSA system.

Note: The order in which H and B are introduced here is perhaps not the most logical which could have been followed. Magnetic induction is normally encountered first, in connection with the force on current-carrying conductors, but the approach followed here is also permissible. It should be mentioned that the above is not intended as a complete, logical treatment of all aspects of electric and magnetic fields, but rather as a reminder to the reader of certain concepts which we need for the discussion of further topics.

We can now define the magnetic flux in a way similar to that given for the electric flux above

$$\Phi = \int_S B_n \, dS \quad \text{weber or volt.second} \qquad (2.25)$$

The question of sign (which also occurs with other surface integrals) may be clarified as follows in the present case (Fig. 2.12). Let us consider an arbitrary surface S, divided up into elements dS, in a magnetic field of magnetic induction B. We determine B_n, the normal component of B for a given element dS. Now B_n is taken as positive if its direction corresponds to the chosen direction of circulation round the closed curve bounding dS, by the corkscrew rule. The direction of circulation chosen is to fit all the elementary areas (see Fig. 2.12), so that the direction of circulation around the curve surrounding the whole surface S is determined uniquely and uniformly. The total magnetic flux through S can now be found from equation (2.25).

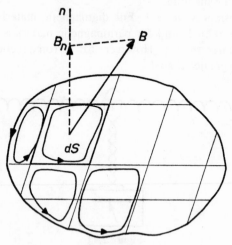

Figure 2.12

Experiment 10

If we determine the total magnetic flux through an arbitrary closed surface in a magnetic field, according to equation (2.25), then we find that the result is zero, even if the imaginary surface encloses a magnet

$$\int_{\substack{\text{closed} \\ \text{surface}}} B_n \, dS = 0 \tag{2.26}$$

It would thus appear that all lines of magnetic flux are closed curves, unlike lines of electric flux, which start from points of positive electric charge and end in points of negative charge.

Experiment 11

Faraday's well-known law of induction states that if the magnetic flux in the region enclosed by a closed conductor changes, this gives rise to an induced e.m.f. in the closed conductor which is equal in magnitude but opposite in sign to the rate of change of flux

$$\text{e.m.f.} = -\frac{d\Phi}{dt} \tag{2.27}$$

This thus gives an electric field E^{ind} due to an induced e.m.f., and not to electric charges.

We have seen that the integral of the electric field strength round a closed curve in an electric field due to stationary charges was zero (equation 2.3). If we integrate the electric field strength round a closed conductor in which an e.m.f. has been induced as above, we find that this integral is equal to the induced e.m.f.

$$\oint E_l^{ind}\, \mathrm{d}l = e_{ind} = -\frac{\mathrm{d}\Phi}{\mathrm{d}t} \qquad (2.28)$$

Now, clearly, we may also write

$$\oint (E_l + E_l^{ind})\, \mathrm{d}l = e_{ind} = -\frac{\mathrm{d}\Phi}{\mathrm{d}t} \qquad (2.29)$$

since, according to equation (2.3), $\oint E_l\, \mathrm{d}l = 0$.

$(E_l + E_l^{ind})$ is the total field strength, so that we may write in general for a closed conductor

$$\oint_{\text{conductor}} E_l\, \mathrm{d}l = -\frac{\mathrm{d}\Phi}{\mathrm{d}t} \qquad (2.30)$$

where E_l now represents the total field strength.

This result is simply another way of writing Faraday's law of induction (equation 2.27).

Maxwell also generalized this law by assuming that every changing magnetic field is surrounded by an electric field whether there is a conductor in the neighbourhood or not, the integral of the electric field strength round an arbitrary closed curve being proportional to the rate of change with time of the magnetic flux enclosed by the curve

$$\oint E_l\, \mathrm{d}l = -\frac{\mathrm{d}\Phi}{\mathrm{d}t} = -\frac{\mathrm{d}}{\mathrm{d}t}\int_S B_n\, \mathrm{d}S \qquad (2.31)$$

where S is the surface defined by the closed curve. The sign convention adopted will be analogous to those described above.

This is Maxwell's second law, and we may, as with Maxwell's first law, draw a similar conclusion:

A changing magnetic field always results in an electric field in and around the region it occupies.

3 MAXWELL'S EQUATIONS

3.1 Maxwell's equations in integral form

The preceding sections have introduced all the basic concepts needed for the treatment of electromagnetic fields. The various ideas, developed more or less independently above, together form the basis for further development. It will not be out of place, therefore, to summarize the most important equations below.

Maxwell's first law: The integral of the magnetic field strength round an arbitrary closed curve is equal to the sum of the conduction currents and displacement currents enclosed by the curve

$$\oint H_l \, dl = i + \frac{d}{dt} \int_S D_n \, dS = i + \frac{d\psi}{dt} \tag{3.1}$$

where

$$i = \int_S J_n \, dS \tag{3.2}$$

J_n is the current density, which is related to the electric field by Ohm's law

$$J = \gamma E \tag{3.3}$$

Maxwell's second law: The integral of the electric field strength round an arbitrary closed curve is equal in magnitude but opposite in sign to the rate of change of the magnetic flux enclosed by the curve

$$\oint E_l \, dl = -\frac{d}{dt} \int_S B_n \, dS = -\frac{d\Phi}{dt} \tag{3.4}$$

Furthermore, the electric flux through a closed surface is equal to the electric charge enclosed by that surface

$$\int_{\substack{\text{closed} \\ \text{surface}}} D_n \, dS = Q \tag{3.5}$$

where

$$Q = \int_V \rho \, dv \tag{3.6}$$

20

ρ is the charge density in the volume V enclosed by S. The magnetic flux through a closed surface is always zero

$$\int_{\substack{\text{closed}\\\text{surface}}} B_n \, dS = 0 \qquad (3.7)$$

The electric flux and magnetic flux are defined as follows

$$D = \varepsilon E = \varepsilon_0 \varepsilon_r E \qquad (3.8)$$

$$B = \mu H = \mu_0 \mu_r H \qquad (3.9)$$

where D has the same direction as E, and B that of H.

Equations (3.1), (3.4), (3.5) and (3.7) are the basic ones, whilst the others may be regarded as connecting equations and definitions.

Let us now consider Maxwell's two laws in connection with the propagation of electromagnetic waves, remembering the conclusions we drew from these laws in the previous section. It should also be noted that Maxwell's laws are independent of one another: the one cannot be derived logically from the other.

As we saw, the first law implies that a changing electric field gives rise to a changing magnetic field, while according to Maxwell's second law this changing magnetic field in its turn produces a changing electric field, and so on. If an electric or magnetic disturbance, i.e. a change in the electric or magnetic field with time, occurs at some point, a chain of events is set into motion. The energy involved continually changes to and from electric and magnetic. This also occurs in a normal low-frequency LC circuit, but in this case the continual changing of the form of electromagnetic energy causes it to be propagated through space, since a changing magnetic field produces an electric field not only at the same point in space but also round that point, whilst the converse is also true.

As an example, an electric current flowing through a wire gives rise to a magnetic field around it; similarly, a dielectric displacement current (due to a changing electric field) gives rise to a magnetic field at and around the site of this current (Fig. 3.1). Summing up, we may state that the energy of the magnetic field induced by an electric fluctuation is not restricted to the region where the original electrical energy was confined, but extends somewhat further. Continuing this argument, we arrive at the conclusion that electromagnetic energy is propagated in wave form.

A disturbance in the electric or magnetic source at a given point does not remain confined to this point, but is propagated outwards until all the energy involved has been dissipated in the losses of the medium in which the process occurs.

Maxwell's equations thus allow a qualitative explanation of the existence and production of electromagnetic waves. We will give a more quantitative explanation below, which will show that electromagnetic waves are propagated with a velocity equal to that of light (3×10^8 m s^{-1}

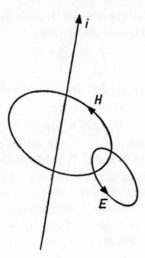

Figure 3.1

in vacuo). In the second half of the last century, Maxwell arrived at a figure which was only slightly lower than this by calculations involving electrostatic and magnetic forces. The wave hypothesis put forward by Maxwell (involving interference, refraction and diffraction phenomena, etc.) was experimentally confirmed twenty years later by Herz.

3.2 Maxwell's equations in differential form

Before applying Maxwell's equations in various situations, we should transform them from the integral form in which they are given in Section 3.1 to a differential form. This is necessary because we have until now represented the quantities E, D, B and H in a slightly misleading form since we are not using vector notation. These four quantities are all vectors, with direction as well as magnitude, and not scalar quantities like heat. In fact, three figures are necessary to define vector quantities completely, while so far we have been using only one symbol with an implied direction. We shall now try, therefore, to formulate Maxwell's equations for all three components of each field quantity, E_x, E_y, E_z and H_x, H_y, H_z, etc., in a rectangular system of coordinates. Such a set of three components fully defines the vector.

For this purpose, we shall apply Maxwell's laws in an infinitesimal parallelepiped with one of its corners at the origin O of a rectangular system of coordinates, and with sides Δx, Δy and Δz (Fig. 3.2). Further details of the nomenclature will be explained with reference to the electric field strength as an example. The electric field strength at the origin can be resolved into the three components E_x, E_y and E_z, as shown in Fig. 3.2. Each component is of course a function of time (t) and position (x, y, z)

$$E_x = E_x(x, y, z, t)$$
$$E_y = E_y(x, y, z, t)$$
etc.

When we differentiate with respect to one of these variables, we must use partial derivatives. We shall use

$$\frac{\partial E_x(x, y, z, t)}{\partial t}$$

to denote differentiation with respect to time at a fixed place (x, y and z constant), while in the operation

$$\frac{\partial E_x(x, y, z, t)}{\partial y}$$

x, z and t are assumed constant.

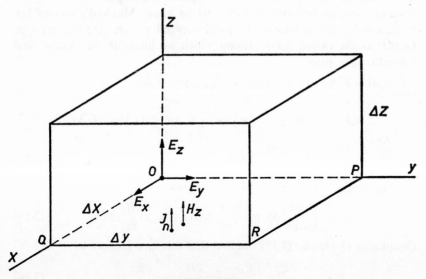

Figure 3.2

We use an infinitesimal parallelepiped in order to ensure that the variation of field strength in the region considered is so small that, for example, at point P (Fig. 3.2) it can be derived from that at O by using only the first term of a Taylor series (this implies that Δx, Δy and Δz are very small).

It will be remembered that a function of one variable can be expanded in a Taylor series as follows

$$f(x+\Delta x) = f(x) + \left(\frac{\partial f(x)}{\partial x}\right)_x \Delta x + \frac{1}{2!}\left(\frac{\partial^2 f(x)}{\partial x^2}\right)_x (\Delta x)^2 + \ldots \quad (3.10)$$

For example, the field strength E_x at the point P may be written

$$E_x(x, y+\Delta y, z, t) = E_x(x, y, z, t) +$$

$$+ \left(\frac{\partial E_x}{\partial y}\right)_{x,y,z,t} \Delta y + \frac{1}{2!}\left(\frac{\partial^2 E_x}{\partial y^2}\right)_{x,y,z,t} (\Delta y)^2 + \ldots \quad (3.11)$$

When Δy is very small, we may neglect second and higher powers of Δy and simply write

$$E_x(P) \approx E_x + \frac{\partial E_x}{\partial y} \cdot \Delta y \quad (3.12)$$

Similarly

$$E_y(Q) \approx E_y + \frac{\partial E_y}{\partial x} \cdot \Delta x \quad (3.13)$$

and so on. E_x and E_y above refer to values at the origin.

After these preliminary remarks, let us apply Maxwell's second law (equation 3.4) to the base of the parallelepiped, i.e. we take the rectangle OQRP as the closed curve around which we integrate the electric field strength. This gives

$$\oint_{OQRP} E_l \, dl = E_x \Delta x + (E_y)_{x+\Delta x}\Delta y - (E_x)_{y+\Delta y}\Delta x - E_y \Delta y$$

$$= E_x\Delta x + \left(E_y + \frac{\partial E_y}{\partial x}\Delta x\right)\Delta y - \left(E_x + \frac{\partial E_x}{\partial y}\Delta y\right)\Delta x - E_y\Delta y$$

$$= \frac{\partial E_y}{\partial x}\Delta x\Delta y - \frac{\partial E_x}{\partial y}\Delta x\Delta y \quad (3.14)$$

The right-hand side of Maxwell's second law is

$$-\int_{OQRP} \frac{\partial B_n}{\partial t} \, dS = -\frac{\partial B_z}{\partial t}\Delta x\Delta y = -\mu\frac{\partial H_z}{\partial t}\Delta x\Delta y \quad (3.15)$$

Combining (3.14) and (3.15), we find that this now reduces to

$$\frac{\partial E_y}{\partial x} - \frac{\partial E_x}{\partial y} = -\mu\frac{\partial H_z}{\partial t} = -\frac{\partial B_z}{\partial t} \quad (3.16)$$

Similarly, we find

$$\frac{\partial E_z}{\partial y} - \frac{\partial E_y}{\partial z} = -\mu\frac{\partial H_x}{\partial t} = -\frac{\partial B_x}{\partial t} \quad (3.17)$$

$$\frac{\partial E_x}{\partial z} - \frac{\partial E_z}{\partial x} = -\mu\frac{\partial H_y}{\partial t} = -\frac{\partial B_y}{\partial t} \quad (3.18)$$

by applying Maxwell's second law to the xOz and yOz planes respectively.

The three equations (3.16 to 3.18) are the differential equivalent of equation (3.4); together they represent another formulation of Maxwell's second law.

For the benefit of those interested in vector notation, it may be mentioned that the vector equivalent of the above three equations is curl $E = -\partial B/\partial t$. This illustrates the advantages of vector notation as regards compactness.

Let us now consider the same parallelepiped, but replace all E's by H's and *vice versa*, and apply Maxwell's first law (equation 3.1), with the current given by equation (3.2). Firstly, we again consider the xOy plane, with OQRP as the closed curve round which the magnetic field strength is integrated. We then find

$$\oint_{\text{OQRP}} H_l \, dl = H_x \Delta x + (H_y)_{x+\Delta x} \Delta y - (H_x)_{y+\Delta y} \Delta x H_y \Delta y$$

$$= \frac{\partial H_y}{\partial x} \Delta x \Delta y - \frac{\partial H_x}{\partial y} \Delta x \Delta y$$

$$= J_z \Delta x \Delta y + \frac{\partial D_z}{\partial t} \Delta x \Delta y$$

or

$$\frac{\partial H_z}{\partial x} - \frac{\partial H_x}{\partial y} = J_z + \frac{\partial D_z}{\partial t} = J_z + \varepsilon \frac{\partial E_z}{\partial t} \tag{3.19}$$

Similarly

$$\frac{\partial H_y}{\partial y} - \frac{\partial H_y}{\partial z} = J_x + \frac{\partial D_x}{\partial t} = J_x + \varepsilon \frac{\partial E_x}{\partial t} \tag{3.20}$$

$$\frac{\partial H_x}{\partial z} - \frac{\partial H_z}{\partial x} = J_y + \frac{\partial D_y}{\partial t} = J_y + \varepsilon \frac{\partial E_y}{\partial t} \tag{3.21}$$

follow from application of Maxwell's first law to the xOz and yOz sides of the parallelepiped. Equations (3.19) to (3.21) represent Maxwell's first law in differential form.

In vector notation, this would be

$$\text{curl } H = J + \frac{\partial D}{\partial t}.$$

For the sake of completeness, we shall now give all the other equations from Section 3.1 in the same differential form.

We assumed above that $\mu H_x = B_x$, etc. This is only true for an isotropic (though possibly inhomogeneous) medium; in other words, we assume that though μ and ε may be functions of position, $\mu = \mu(x, y, z)$ and $\varepsilon = \varepsilon(x, y, z)$, the properties of the material at a given point are the same in all directions.

Ohm's law in component notation, under the same assumptions, is then

$$J_x = \gamma E_x$$
$$J_y = \gamma E_y \qquad (3.22$$
$$J_z = \gamma E_z$$

Application of equations (3.5) and (3.6), relating to the electric flux, to the parallelepiped gives

$$\oint D_n \, dS = \int_V \rho(x, y, z) \, dv = -D_z \Delta x \Delta y + (D_z)_{z+\Delta z} \Delta x \Delta y - D_y \Delta x \Delta z$$

$$+ (D_y)_{y+\Delta y} \Delta x \Delta z - D_x \Delta y \Delta z + (D_x)_{x+\Delta x} \Delta y \Delta z$$

$$= \left(\frac{\partial D_x}{\partial x} + \frac{\partial D_y}{\partial y} + \frac{\partial D_z}{\partial z} \right) \Delta x \Delta y \Delta z = \rho(x, y, z) \Delta x \Delta y \Delta z$$

or

$$\left(\frac{\partial D_x}{\partial x} + \frac{\partial D_y}{\partial y} + \frac{\partial D_z}{\partial z} \right) = \rho \qquad (3.23)$$

Similarly, equation (3.7) gives

$$\frac{\partial B_x}{\partial x} + \frac{\partial B_y}{\partial y} + \frac{\partial B_z}{\partial z} = 0 \qquad (3.24)$$

which represents the description of the behaviour of magnetic flux in differential notation.

One commonly used equation, the 'continuity equation', is derived from the above equations by differentiating equations (3.20), (3.21) and (3.19) with respect to x, y and z respectively, and adding the results. With equation (3.23), we then find

$$\frac{\partial J_x}{\partial x} + \frac{\partial J_y}{\partial y} + \frac{\partial J_z}{\partial z} = -\frac{\partial \rho}{\partial t} \qquad (3.25)$$

The sets of partial differential equations derived in this section form the basis of the study of the electromagnetic field. The solutions of these equations will of course contain a number of arbitrary constants, which must be given such values as conform with the boundary conditions. These boundary conditions may be regarded as restrictions caused by the presence of conducting surfaces, dielectric interfaces, energy sources and the like at various places, as a result of which the value of certain field quantities is prescribed at these places.

The solution of Maxwell's equations is not in general a simple matter; it will often be necessary to use approximations and to make various simplifying assumptions before a solution can be obtained.

4 BOUNDARY CONDITIONS FOR ELECTROMAGNETIC FIELDS

When the equations developed in Section 3.2 are applied to practical situations, the region considered nearly always contains one or more air–metal, air–dielectric or other interfaces. Such interfaces often form the boundary of the region, and the field quantities are usually given along these boundaries. The solutions of the foregoing equations must of course be adjusted to coincide with the values given along the boundaries.

We shall now derive a number of general conditions for interfaces by applying Maxwell's equations in integral form.

4.1 Interfaces between dielectrics

Let us consider the interface between two dielectrics 1 and 2, with dielectric constants ε_1 and ε_2 and permeabilities μ_1 and μ_2 respectively (Fig. 4.1).

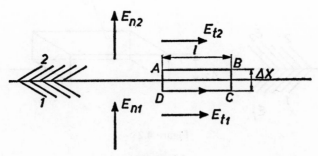

Figure 4.1

We resolve the electric field strengths E_1 and E_2 in the two media near the interface into a tangential and a normal component as shown in the figure, and apply Maxwell's second law to the closed curve $ABCD$,

27

a rectangle of sides l and Δx which is symmetrical with respect to the interface. Now if we let Δx tend to zero, the area of the rectangle will also tend to zero, until the closed curve $ABCD$ will enclose no magnetic flux. Application of equation (3.4) in the limit as $\Delta x \to 0$ thus gives

$$E_{t1}\,l - E_{t2}\,l = 0$$

or
$$E_{t1} = E_{t2} \tag{4.1}$$

Now since
$$D_{t1} = \varepsilon_1 E_{t1} \quad \text{and} \quad D_{t2} = \varepsilon_2 E_{t2}$$

it follows from equation (4.1) that

$$D_{t1} = \frac{\varepsilon_1}{\varepsilon_2} D_{t2} \tag{4.2}$$

The tangential component of the electric field strength is thus continuous at the interface between the two media (i.e., has the same value on both sides of the interface), while the tangential components of the dielectric displacements on the two sides of the interface are in equal ratio to the dielectric constants.

We now apply equation (3.5), concerning the enclosed electric charge, to a flat box symmetrical with respect to the interface, as sketched in Fig. 4.2. We again let the thickness Δx of the box tend to zero. Now if the normal component of the dielectric displacement at the interface is written D_n, we find

$$-D_{n1}S + D_{n2}S = \sigma S$$

or
$$-D_{n1} + D_{n2} = \sigma \tag{4.3}$$

where σ is the surface charge density in C m^{-2} at the interface.

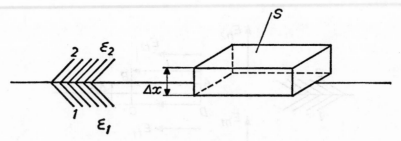

Figure 4.2

If no free charge is present at the surface, we see that the normal component of the dielectric displacement is continuous at the interface between the two dielectrics, since equation (4.3) then gives

$$D_{n1} = D_{n2} \tag{4.4}$$

The normal components of the electric field strength are then related by

$$E_{n1} = \frac{\varepsilon_2}{\varepsilon_1} E_{n2} \qquad (4.5)$$

i.e. they are inversely proportional to the dielectric constants.

It will be clear from equation (4.5) that when the field strength is high, cracks in the material perpendicular to the lines of force greatly increase the risk of breakdown.

It follows from the above that the lines of force of an electric field will be refracted where the electric field passes from one medium to

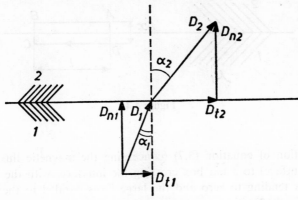

Figure 4.3

another (Fig. 4.3). If there is no free charge at the interface, then it follows from the figure and from equations (4.2) and (4.4) that

$$\frac{\tan \alpha_1}{\tan \alpha_2} = \frac{D_{t1}}{D_{n1}} \cdot \frac{D_{n2}}{D_{t2}} = \frac{\varepsilon_1}{\varepsilon_2} = \frac{\varepsilon_{r1}}{\varepsilon_{r2}} \qquad (4.6)$$

We shall now proceed to derive similar relationships for magnetic field quantities at the boundary between two dielectrics, with the aid of Maxwell's first law.

We integrate the magnetic field strength round the rectangle $ABCD$ (Fig. 4.4), and again let the width Δx tend to zero. In the limit, this gives

$$H_{t1} . l - H_{t2} . l = sl$$

or

$$H_{t1} - H_{t2} = s \qquad (4.7)$$

where s is the current density at the interface in A m^{-1}; the contribution of the dielectric displacement tends to zero with Δx.

If there is no current at the interface ($s = 0$), then equation (4.7) becomes

$$H_{t1} = H_{t2} \qquad (4.8)$$

2*

and consequently

$$B_{t1} = \frac{\mu_1}{\mu_2} B_{t2} \qquad (4.9)$$

When there is no current at the interface, therefore, the tangential component of the magnetic field strength is continuous at the interface, while the ratio of the tangential components of the magnetic induction equals that of the permeabilities of the two media.

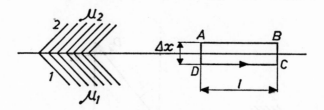

Figure 4.4

Application of equation (3.7) (concerning the magnetic flux through a closed surface) to a flat box crossing the interface, with the thickness of the box tending to zero and the large faces parallel to the interface (Fig. 4.2) in the limiting case gives

$$B_{n1} = B_{n2} \qquad (4.10)$$

and hence

$$H_{n1} = \frac{\mu_2}{\mu_1} H_{n2} \qquad (4.11)$$

In other words, the normal component of the magnetic induction is continuous at the interface of two media, while the ratio of the normal components of the magnetic field strength is inversely proportional to that of the permeabilities.

The lines of magnetic induction suffer refraction at the boundary between two dielectrics (Fig. 4.5), according to the equation

$$\frac{\tan \alpha_1}{\tan \alpha_2} = \frac{B_{t1}}{B_{n1}} \cdot \frac{B_{n2}}{B_{t2}} = \frac{\mu_1}{\mu_2} = \frac{\mu_{r1}}{\mu_{r2}} \qquad (4.12)$$

This follows from the figure and equations (4.9) and (4.10).

If, for example, $\mu_{r2} \gg \mu_{r1}$, which is the case when medium 1 is air and medium 2 is iron, the lines of magnetic induction leave the iron nearly at right angles.

Summing up, we may state that the following relations hold at the interface between two dielectrics, in the absence of free charge and surface current

$$E_{t1} = E_{t2} \rightarrow D_{t1} = \frac{\varepsilon_1}{\varepsilon_2} D_{t2}; \qquad H_{t1} = H_{t2} \rightarrow B_{t1} = \frac{\mu_1}{\mu_2} B_{t2}$$

$$D_{n1} = D_{n2} \rightarrow E_{n1} = \frac{\varepsilon_2}{\varepsilon_1} E_{n2}; \qquad B_{n1} = B_{n2} \rightarrow H_{n1} = \frac{\mu_2}{\mu_1} H_{n2}$$

(4.13)

If there are free charges (σ in C m^{-2}) or currents (s in A m^{-1}) at the interface, we find

$$E_{t1} = E_{t2} \rightarrow D_{t1} = \frac{\varepsilon_1}{\varepsilon_2} D_{t2}; \qquad H_{t1} - H_{t2} = s$$

$$-D_{n1} + D_{n2} = \sigma; \qquad B_{n1} = B_{n2} \rightarrow H_{n1} = \frac{\mu_2}{\mu_1} H_{n2}$$

(4.14)

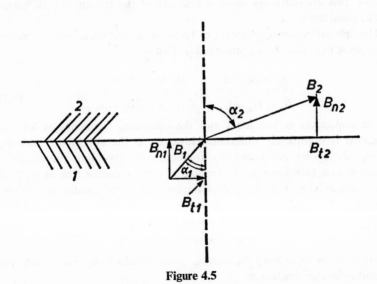

Figure 4.5

4.2 Interface between dielectrics and ideal conductors

If one of the two media in question is a perfect conductor ($\gamma \rightarrow \infty$), then at the interface

$$E_t = B_n = 0 \tag{4.15}$$

As we saw in Section 4.1, these quantities are continuous at the interface between any two media; and no electric or magnetic field can exist in a perfect conductor—for if any field, no matter how small, did exist it would, according to Ohm's law with $\gamma \to \infty$, give rise to an infinite current which would compensate the field. It follows further from Maxwell's laws that no magnetic field can exist, because a magnetic field is always associated with the production of an electric field and we have just seen that an electric field cannot exist. Currents and charges can thus only exist at the surface of a perfect conductor, not in its interior.

The normal component of the dielectric displacement is of course also zero in the ideal conductor, but not outside it. D_n thus exhibits a discontinuity at the surface. According to equation (4.3), this discontinuity in D_n is a measure of the surface charge on the conductor; if medium 1 in Fig. 4.3 is the conductor, then $D_{n1} = 0$ and $D_{n2} \neq 0$, or

$$D_{n2} = \sigma \tag{4.16}$$

Similarly, the tangential component of the magnetic field strength exhibits a discontinuity at the interface between a dielectric and an ideal conductor, this discontinuity being a measure of the current on the surface of the conductor.

The situation at the interface between a dielectric and a perfect conductor may thus be summarized as follows

$$E_t = D_t = 0; \qquad D_n = \sigma \to E_n = \frac{\sigma}{\varepsilon}$$
$$B_n = H_n = 0; \qquad H_t = s \to B_t = \mu s \tag{4.17}$$

If the conductor is not ideal and the operating frequency is low, the situation is completely different. However, if the conductor, although non-ideal, is very good and the frequency is very high, the current is confined to a thin layer at the surface of the conductor (the skin effect). The *penetration or skin depth* δ of the current in the conductor, which is defined as the distance from the surface at which the amplitude of the current falls to $1/e$ times that at the surface, is given by

$$\delta = (1/\pi f \gamma \mu)^{\frac{1}{2}} \quad \text{metres} \tag{4.18}$$

where f is the operating frequency, γ the conductivity and μ the permeability of the conductor.

5 ELECTROMAGNETIC WAVES IN UNBOUNDED DIELECTRICS

5.1 The wave equation for plane waves

In this section we shall consider Maxwell's equations for an unbounded dielectric in further detail; this may be regarded as a continuation of the qualitative introduction to wave phenomena given above. We shall assume that no currents flow in the dielectric, i.e. that the conductivity of the medium

$$\gamma = 0 \tag{5.1}$$

The results obtained will then be directly applicable to air or vacuum if ε is replaced by ε_0 and μ by μ_0.

At this time there would be little point in carrying out complicated calculations on the most general case imaginable of waves in an unbounded medium. In order to obtain a clearer picture of the situation, we shall confine ourselves, apart from the restriction mentioned above in equation (5.1), to simple cases in which we limit calculation to the minimum required to bring out the important features.

Now it will be clear that the situation can be simplified in many different ways. For example, we could assume, apart from equation (5.1), that all components of E and H, which are in general functions of time and all three positional co-ordinates, are functions of time and one positional co-ordinate only, e.g. z. This would give

$$\frac{\partial E_x}{\partial x} = \frac{\partial E_x}{\partial y} = \frac{\partial E_y}{\partial x} = \frac{\partial E_y}{\partial y} = \frac{\partial E_z}{\partial x} = \frac{\partial E_z}{\partial y} = 0 \tag{5.2}$$

and with a similar condition for the components of H. Equation (5.1), together with Ohm's law, would give

$$J_x = J_y = J_z = 0 \tag{5.3}$$

However, we shall not use these simplifying assumptions here; it may be left as an exercise to the reader to work out their implications.

The situation we shall consider is even simpler. We assume that the electric field has only one component, the y component E_y. Hence

$$E_x = E_z = 0 \tag{5.4}$$

We assume E_y to be a function of x and t, but not of the two other positional co-ordinates y and z

$$E_y = E_y(x) \quad \text{and} \quad \frac{\partial E_y}{\partial y} = \frac{\partial E_y}{\partial z} = 0 \tag{5.5}$$

Further, of course, we assume equation (5.1).

Components of field quantities which are constant and independent of time do not interest us here. If they occur, we shall assume them to be zero.

If we now substitute equations (5.1) and (5.3) to (5.5), representing the assumptions and simplifications we have made, into Maxwell's equations as given in Section 3.2, we find

$$\frac{\partial B_x}{\partial t} = 0 \quad \text{and} \quad \frac{\partial B_y}{\partial t} = 0 \tag{5.6}$$

$$\frac{\partial E_y}{\partial x} = -\frac{\partial B_z}{\partial t} \tag{5.7}$$

$$-\frac{\partial H_z}{\partial x} = \frac{\partial D_y}{\partial t} \tag{5.8}$$

It follows from equation (5.6) that B_x and B_y are constant; we shall assume them to be zero, as we are not interested in them.

Differentiation of equation (5.7) with respect to x and equation (5.8) with respect to t gives

$$\frac{\partial^2 E_y}{\partial x^2} = -\frac{\partial^2 B_z}{\partial x \, \partial t} \quad \text{and} \quad \frac{\partial^2 H_z}{\partial x \, \partial t} = -\frac{\partial^2 D_y}{\partial t^2} \tag{5.9}$$

If we now assume that the medium in question is isotropic (i.e. that its properties do not depend on direction) and homogeneous (i.e. that ε and μ are not functions of position), then it follows from equation (5.9) that

$$\frac{\partial^2 E_y}{\partial x^2} = \varepsilon\mu \frac{\partial^2 E_y}{\partial t^2} \tag{5.10}$$

If we had differentiated equation (5.7) with respect to t and equation (5.8) with respect to x, we would have found

$$\frac{\partial^2 H_z}{\partial x^2} = \varepsilon\mu \frac{\partial^2 H_z}{\partial t^2} \tag{5.11}$$

In the simplest case, then, where the electric field has only one component E_y, and the magnetic field also only has one component H_z, both these quantities being functions of x and t only, we find the same equation for both components (equations 5.10 or 5.11). This may be called the *wave equation,* for it represents a travelling wave, as we shall see below.

If we had not made the above simplifying assumptions, this line of argument would have led to a wave equation for each of the six components of the electromagnetic field, and each equation would have contained more terms because of the dependence of the variables on y and z as well. These equations would, in fact, have had the general form

$$\frac{\partial^2 A}{\partial x^2} + \frac{\partial^2 A}{\partial y^2} + \frac{\partial^2 A}{\partial z^2} = \varepsilon\mu \frac{\partial^2 A}{\partial t^2} \tag{5.12}$$

where A represents E_x, E_y, E_z, H_x, H_y or H_z.

The general solution of the simplified wave equation derived above (such as equation 5.10) is

$$E_y = f_1(t - x\sqrt{\varepsilon\mu}) + f_2(t + x\sqrt{\varepsilon\mu}) \tag{5.13}$$

as can easily be verified by substitution in the wave equation. In equation (5.13), f_1 and f_2 represent arbitrary functions of their respective arguments, $(t - x\sqrt{\varepsilon\mu})$ and $(t + x\sqrt{\varepsilon\mu})$. The precise form of these functions is determined by: (a) the form of excitation (e.g. a sinusoidal wave source); (b) the boundary conditions in a homogeneous medium bounded by certain surfaces or interfaces (as in a waveguide).

Let us assume for the sake of simplicity that $f_2 \equiv 0$. For the moment, we are only interested in solutions of the wave equation which can be written as $f_1(t - x\sqrt{\varepsilon\mu})$, i.e. electric fields of the same form

$$E_y = f_1(t - x\sqrt{\varepsilon\mu}) \tag{5.14}$$

Now what type of function is this, and how does it vary with x and t?

Let us suppose that at a given moment $t = t_1$ the function varies with x as shown in Fig. 5.1. If we now consider the situation at a later instant

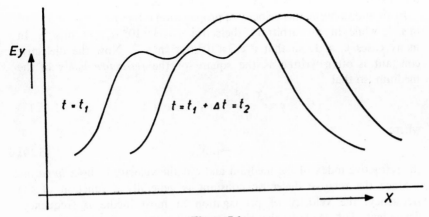

Figure 5.1

of time $t_2 > t_1$, and keep the $(t - x\sqrt{\varepsilon\mu})$ constant, the function will still have the same value. This means that x must change too, so that the same function $E_y(x)$ is obtained for a slightly different value of x at this later time t_2. In this case

$$t - x\sqrt{\varepsilon\mu} = \text{constant}$$

or

$$x = \frac{t}{\sqrt{\varepsilon\mu}} + a \tag{5.15}$$

where a is a constant.

In other words, when t increases by Δt, if we want E_y to keep the same value, we must increase x by

$$\Delta x = \frac{\Delta t}{\sqrt{\varepsilon\mu}} \tag{5.16}$$

Now let us introduce a new variable

$$v = 1/\sqrt{\varepsilon\mu} \tag{5.17}$$

This v has the dimension of velocity.

In fact, as time passes the pattern shown in Fig. 5.1 will move to the right along the x axis with a velocity v, given by equation (5.17). As our qualitative arguments had indicated already, Maxwell's equations thus imply the existence of *electromagnetic waves*, in this case a travelling wave propagated along the x axis to the right with a velocity $v = 1/(\sqrt{\varepsilon\mu})$. This is the significance of the function $f_1(t - x\sqrt{\varepsilon\mu})$.

Similarly, $f_2(t + x\sqrt{\varepsilon\mu})$ represents a wave travelling along the x axis with the same velocity to the left.

The general solution of the wave equation is thus compatible with the existence of two superimposed travelling waves, moving in opposite directions with a velocity v given by equation (5.17).

In air and vacuum, this velocity of propagation is $v = 1/(\varepsilon_0\mu_0)^{\frac{1}{2}} = 3 \times 10^8$ m s^{-1}, while in an arbitrary dielectric $v = 3 \times 10^8/(\varepsilon_r\mu_r)^{\frac{1}{2}}$ m s^{-1}. In many cases $\mu_r \approx 1$, so that $v \approx 3 \times 10^8/(\varepsilon_r)^{\frac{1}{2}}$ m s^{-1}. Now the dielectric constant is often written as the square of the *refractive index* of the medium, so that

$$v \approx \frac{3 . 10^8}{n} \quad \text{m s}^{-1} = c/n \tag{5.18}$$

where

$$n = (\varepsilon_r)^{\frac{1}{2}} \tag{5.19}$$

the refractive index of the medium and c is the velocity of light *in vacuo*.

Since the relative dielectric constant is generally a function of the frequency, the velocity of propagation in most media is frequency-dependent. This leads to the phenomenon known as *dispersion*: when a frequency band of finite width is transmitted through the medium, the different frequencies will arrive at different times.

In the case we are considering (see Fig. 5.2), the direction of propagation is perpendicular to the plane containing the components of the field, E_y and H_z, and the positive direction of v is given by the corkscrew rule for rotation from E_y to H_z through the smallest possible angle. This rule can be generalized to the case where more components are present. The direction of propagation of the wave is then given by the corkscrew rule for rotation from the resultant electric field to the resultant magnetic field through the smallest possible angle.

We shall now proceed to specify the situation somewhat more closely by assuming that the electromagnetic waves are excited *harmonically*, by which we mean that the wave source produces sine-wave or cosine-wave disturbances (for example, by means of a vertically oscillating

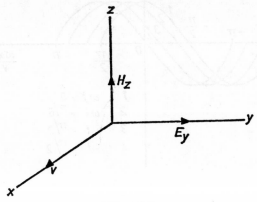

Figure 5.2

dipole). In this case, the function $f(t - x/v)$ will have a more closely defined form. In fact, the electrical field strength may now be written

$$E_y = E_m \cos \omega \left(t - \frac{x}{v} \right)$$
$$E_x = E_z = 0$$

(5.20)

where $\omega = 2\pi f$, and E_m is the amplitude of the electric field strength. Mathematically, E_m and ω are arbitrary constants. The fact that we could assume the validity of equation (5.20) if excitation was harmonic follows from the above reasoning, which showed that the form of the function is maintained during propagation, i.e. if the disturbance starts off in a harmonic form (a cosine wave), then it will be propagated in the same form.

It goes without saying that equation (5.20) is a solution of the wave equation, for it is only a special case of the more general solution in equation (5.13).

Fig. 5.3 shows one period of the wave in equation (5.20), the wave form being shown for four different values of ωt. If the wave were visible, these four curves would be 'stills' of the electric field at the stated times. The wave is propagated from left to right along the x axis, its general form being preserved. This wave is called a *transverse wave*, because the field only has components at right angles to the direction of propagation.

We find further that, in accordance with the assumptions we made at the start of the calculations, the electric field is uniform in a plane

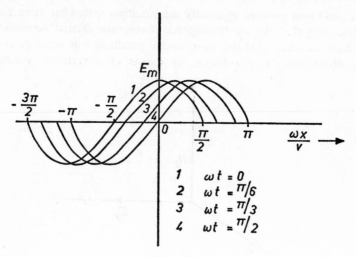

1	$\omega t = 0$
2	$\omega t = \pi/6$
3	$\omega t = \pi/3$
4	$\omega t = \pi/2$

Figure 5.3

perpendicular to the direction of propagation. At a given moment, the electric field has the same value at all points of a given plane at right angles to the direction of propagation. In the plane $x = x_0$, the electric field strength varies with time according to the equation

$$E_y = E_m \cos \omega \left(t - \frac{x_0}{v} \right) \qquad (5.21)$$

We now determine the associated magnetic field by substituting the solution (equation 5.20) of the wave equation in equation (5.8). This gives

$$\frac{\partial H_z}{\partial x} = -\frac{\partial E_y}{\partial t} = \varepsilon E_m \omega \sin \omega \left(t - \frac{x}{v} \right) \qquad (5.22)$$

Integration with respect to x gives

$$H_z = \varepsilon E_m \frac{\omega}{\omega} v \cos \omega \left(t - \frac{x}{v} \right) = \sqrt{\frac{\varepsilon}{\mu}} E_m \cos \omega \left(t - \frac{x}{v} \right) \qquad (5.23)$$

The following notation is generally used

$$\beta = \frac{\omega}{v} = \frac{2\pi f}{v}; \quad \lambda = \frac{v}{f} = \frac{2\pi}{\beta} \tag{5.24}$$

In the present case, then, the expressions for the electromagnetic field may be written

$$E_y = E_m \cos(\omega t - \beta x) \tag{5.25}$$

$$H_z = \sqrt{\frac{\varepsilon}{\mu}} \, E_m \cos(\omega t - \beta x) = \sqrt{\frac{\varepsilon}{\mu}} \, E_y \tag{5.26}$$

These represent an electromagnetic wave propagated to the right along the x axis with a velocity $v = 1/\sqrt{\varepsilon \mu}$.

In the simple case we have considered here, the wave is a *plane wave:* E and H are uniform in a plane perpendicular to the direction of propagation. In a plane of this type, E and H vary with time, while if we should follow one of the planes with velocity v, we would find that E and H were constant.

It follows further from equations (5.25) and (5.26) that E and H are in phase in a travelling plane wave (their ratio is a constant real number). The ratio of E_y to H_z is called the *intrinsic impedance* of the medium

$$\eta = \frac{E_y}{H_z} = \sqrt{\frac{\mu}{\varepsilon}} \tag{5.27}$$

In vacuum or air, $\eta = \sqrt{\dfrac{\mu_0}{\varepsilon_0}} = 120\pi$.

The situation is summarized in Fig. 5.4. Each arrow gives the value of E or H in the plane in question at a given moment; the fields vary

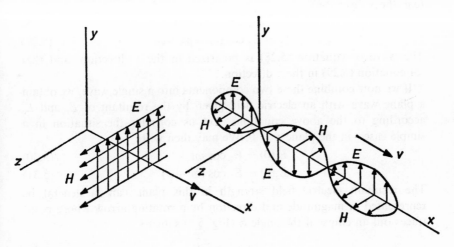

Figure 5.4

sinusoidally along the x axis, and the whole pattern moves along the axis with velocity v. At any given point, E and H vary sinusoidally with time.

5.2 Polarization of waves

The travelling plane transverse wave described in the above section is also polarized. What does this mean? An electromagnetic wave is said to be linearly polarized if the resultant electric field strength always points in the same direction. It is true that the magnitude of the field varies harmonically, but its direction does not change (apart from the change in sign). The restriction $E_x = E_z = 0$ imposed on the wave makes it polarized; together with the other condition, $\partial E_y/\partial y = \partial E_y/\partial z = 0$, this makes the wave a plane polarized one.

In an arbitrary, unpolarized wave, E and H vary both in magnitude and direction. In general, an unpolarized plane wave can be regarded as a number of plane polarized waves. The component waves can be polarized in any direction, and indeed the components of the system can have any amplitude and phase. If all components have the same orientation, the wave is said to be linearly polarized. The direction of polarization, unlike the case in optics, is taken as the direction of the electric field.

Such a superposition principle is possible in this case because the wave equation is linear: a linear combination of two independent solutions is also a solution.

Let us now consider two uniform plane waves of different amplitude, phase and direction of polarization, but propagated in the same direction (e.g. the x direction)

$$E_y = E_1 \cos (\omega t - \beta x) \tag{5.28}$$
$$E_z = E_2 \cos (\omega t - \beta x + \varphi) \tag{5.29}$$

The wave of equation (5.28) is polarized in the y direction, and that of equation (5.29) in the z direction.

If we now combine these two components into a single wave, we obtain a plane wave with an electric field given by the resultant of E_y and E_z according to the above equations. Let us consider the situation in a simple case—in the plane $x = 0$. We may then write

$$E_y(0) = E_1 \cos \omega t \tag{5.30}$$
$$E_z(0) = E_2 \cos (\omega t + \varphi) \tag{5.31}$$

The resultant electric field strength in this plane can in general be represented in magnitude and direction by a rotating arrow whose point traces out an ellipse if the angle φ (Fig. 5.5) satisfies

$$\varphi = \pm \frac{\pi}{2} \tag{5.32}$$

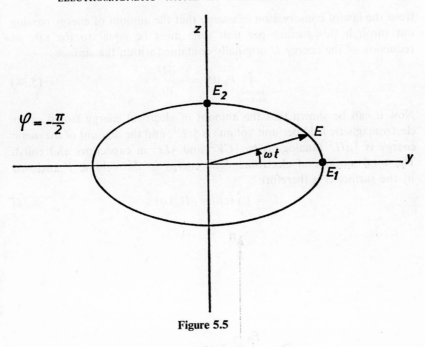

Figure 5.5

Such a wave is said to be *elliptically* polarized.

If

$$\varphi = \frac{\pi}{2} \quad \text{and} \quad E_1 = E_2 \tag{5.33}$$

the resultant electric field may be represented by a moving arrow whose tip traces out a circle; the wave is then said to be *circularly* polarized.

5.3 The Poynting vector

We shall now define a measure for the amount of energy propagated in an electromagnetic wave. We define the intensity P of electromagnetic radiation as the energy passing per unit time (one second in our system) through unit area (m²) at right angles to the direction of propagation. This intensity P is a vector quantity: it has both a magnitude and a direction.

Now let us consider an element dS of a surface S (Fig. 5.6): n is the normal to dS; P is the intensity vector or *Poynting vector*; P_n is the component of P normal to dS; then $P_n . dS$ is the amount of energy passing through dS per second. The amount of energy passing through the whole surface S per second is given by $\int_s P_n \, dS$. If S is a closed surface, it follows from the law of conservation of energy that the amount of energy passing out through this surface per unit time must be equal to the rate of reduction of the energy U originally contained within the surface

$$\int_{\substack{\text{closed}\\ \text{surface}}} P_n \, dS = -\frac{dU}{dt} \tag{5.34}$$

Now it can be shown that the amount of electrical energy stored in an electromagnetic field per unit volume is $\frac{1}{2}\varepsilon E^2$, and the amount of magnetic energy is $\frac{1}{2}\mu H^2$ (analogous to $\frac{1}{2}CV^2$ and $\frac{1}{2}Li^2$ in capacitors and coils). The total amount of electromagnetic energy in the volume V enclosed by the surface S is therefore

$$U = \frac{1}{2} \int_V (\varepsilon E^2 + \mu H^2) \, dV \tag{5.35}$$

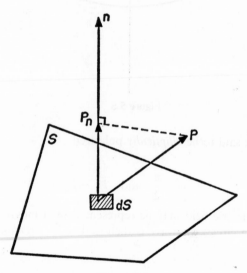

Figure 5.6

Substitution in (5.34) gives

$$\int_{\substack{\text{closed}\\ \text{surface}}} P_n \, dS = -\frac{1}{2}\frac{\partial}{\partial t} \int_V (\varepsilon E^2 + \mu H^2) \, dV$$

$$= -\int_V \left(E\frac{\partial D}{\partial t} + H\frac{\partial B}{\partial t} \right) dV \tag{5.36}$$

Let us now apply this equation to the linearly polarized plane wave discussed in the previous section, taking as volume the parallelepiped of side dx, dy and dz as shown in Fig. 5.7. Since the electromagnetic wave only has the components E_y and H_z and is propagated in the x direction, we need only consider the faces dy, dz of the parallelepiped, since no energy passes through the other faces; we then have $P_n = P_x = P$. If we now expand the energy flow in a Taylor series, taking only the first two terms as we did above in the derivation of Maxwell's equations, application of equation (5.36) to our volume element gives

$$-(P_x \, dy \, dz)_x + (P_x \, dy \, dz)_{x+dx} \approx \frac{\partial P_x}{\partial x} \, dx \, dy \, dz$$

$$= -\left(E \frac{\partial D}{\partial t} + H \frac{\partial B}{\partial t}\right) dx \, dy \, dz$$

$$= \left(E_y \frac{\partial H_z}{\partial x} + H_z \frac{\partial E_y}{\partial x}\right) dx \, dy \, dz$$

$$= \left\{\frac{\partial}{\partial x}(E_y H_z)\right\} dx \, dy \, dz \qquad (5.37)$$

In this derivation we made use of equations (3.16) and (3.21) for the special case of the plane wave considered here, with only E_y and H_z differing from zero.

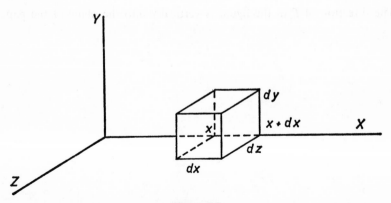

Figure 5.7

Integrating equation (5.37) we find that in our case the energy flow is given by

$$P_x = E_y H_z \qquad (5.38)$$

We have already seen that the direction of energy flow was the x direction, the direction at right angles to E_y and H_z, and according to the corkscrew rule to rotation from E_y to H_z through the smallest possible angle.

This rule, derived for the special case considered here, can be generalized. If the electromagnetic field has several components, the direction of energy flow for a plane electromagnetic wave propagated in an arbitrary direction

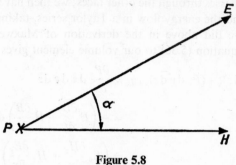

Figure 5.8

will be at right angles to the resultant E and H, in the direction given
by the corkscrew rule for rotation from E to H through the smallest
possible angle. The magnitude of the intensity vector in this case will be
given by the modulus of E times the modulus of H times the sine of the
angle between E and H (Fig. 5.8)

$$P = EH \sin \alpha \tag{5.39}$$

The direction of P in the figure is vertically into the plane of the paper.

6 ELECTROMAGNETIC WAVES IN CONDUCTING MEDIA

So far, we have only solved the wave equation for the special case of an ideal dielectric of zero conductivity γ. We shall now extend our considerations to media of finite conductivity.

If we were to assume a very general time dependence of the electromagnetic field, the calculations involved in the solution of the wave equation would be quite complicated. It is considerably simpler only to consider a harmonic time dependence. This assumption that the components of the electromagnetic field vary sinusoidally with time might appear to limit our scope drastically. However, the reader will remember that most of the functions we are dealing with are periodic, if otherwise arbitrary in form, and can hence be resolved into a number of harmonic components by Fourier analysis. The solution of the wave equation can be written for each of these components (once we have completed the considerations given here); and the sum of these component solutions gives the overall solution, since Maxwell's equations are linear, as we have already mentioned. Our restriction of the field to harmonic functions is thus not as drastic as it might seem. We shall therefore assume from now on that E and H vary harmonically with time, and can thus be written in the form

$$E(x, y, z, t) = E_0(x, y, z) \cos (\omega t + \varphi)$$

However, in order to simplify the calculations and to conform with current usage, we shall make use of a complex notation for the components of the field.

6.1 Complex notation

The definitions and rules of calculation for complex numbers are summarized in an appendix; we shall assume here that they are known.

On the basis of Euler's equation (see appendix), we can write the following identity

$$A \cos (\omega t + \varphi) = \frac{A}{2} \{e^{i(\omega t + \varphi)} + e^{-i(\omega t + \varphi)}\} \qquad (6.1)$$

The right-hand side of this equation is simply another way of writing the left-hand side; this relation is generally applicable.

In electrical theory, however, we do not generally make use of complex notation in this way, but base our considerations on the equation

$$e^{i(\omega t + \varphi)} = \cos (\omega t + \varphi) + i \sin (\omega t + \varphi) \qquad (6.2)$$

(also due to Euler). It follows from this that

$$\text{Re } e^{i(\omega t + \varphi)} = \cos (\omega t + \varphi) \qquad (6.3)$$

where Re signifies 'the real part of'.

Equation (6.3) is also an identity, and its use cannot lead to error if we write $\cos (\omega t \times \varphi)$ as on the left-hand side. However, the convention has been adopted that the Re should be omitted, so that when we write $e^{i(\omega t + \phi)}$ we mean $\cos (\omega t \times \varphi)$, or rather Re $e^{i(\omega t + \phi)}$. Now this can easily lead to errors in expressions involving products of such functions. For example, if we write

$$e^{i(\omega t + \varphi)} \, e^{i(\omega t + \psi)}$$

we mean

$$\text{Re } \{e^{i(\omega t + \varphi)}\} \, \{\text{Re } e^{i(\omega t + \psi)}\} = \cos (\omega t + \varphi)\text{c os} (\omega t + \psi)$$

and definitely not

$$e^{i(\omega t + \varphi)} \, e^{i(\omega t + \psi)} = e^{i(2\omega t + \varphi + \psi)}$$

since

$$\text{Re } \{e^{i(2\omega t + \varphi + \psi)}\} \neq \cos (\omega t + \phi) \cos (\omega t + \psi)$$

In such cases, it is really more sensible to make use of equation (6.1).

On this understanding, we shall now use the notation of equation (6.3) in the normal way to represent an electric field varying harmonically with time according to the equation

$$E(x, y, z, t) = E_0^1(x, y, z) \cos (\omega t + \varphi) \qquad (6.4)$$

We thus write this field

$$E_0^1(x, y, z) \, e^{j(\omega t + \varphi)} \qquad (6.5)$$

In fact, this field is often written in the alternative form

$$E_0^1(x, y, z) \, e^{j(\omega t + \varphi)} = E_0^1(x, y, z) \, e^{j\varphi} \, e^{j\omega t} = E_0(x, y, z) \, e^{j\omega t} \qquad (6.6)$$

where $E_0^1(x, y, z)$ represents a real amplitude and $E_0(x, y, z)$ a 'complex amplitude'. In the latter case, the phase factor is contained in the position-dependent term.

Equation (6.6) thus implies the correspondence

$$E(x, y, z, t) = \text{Re } \{E_0(x, y, z) \, e^{j\omega t}\}$$

where j represents the imaginary unit. (This is often used in electrical theory, to avoid confusion with the current i.)

For example, in Section 5.1 we represented a plane wave travelling along the x axis in the direction of increasing values of x by $E_m \cos(\omega t - \beta x + \varphi)$. In our new notation, this becomes

$$E_m \, e^{j(\omega t - \beta x + \varphi)} = E_m \, e^{j\varphi} \, e^{j(\omega t - \beta x)} = E_m^1 \, e^{j\omega t} \, e^{-j\beta x} \qquad (6.7)$$

where we again write E_m^1 for $E_m \, e^{j\phi}$ (in fact, we have now introduced the extra phase φ).

Now there is one further question which needs to be considered. So far, we have used Euler's equation in the form

$$\cos(\omega t + \varphi) = \operatorname{Re} \{ e^{j(\omega t + \varphi)} \} = \operatorname{Re} \{ e^{\sqrt{-1}(\omega t + \varphi)} \} \qquad (6.8)$$

However, according to Euler we could equally well have written

$$\cos(\omega t + \varphi) = \operatorname{Re} \{ e^{-\sqrt{-1}(\omega t + \varphi)} \} \qquad (6.9)$$

since

$$e^{-\sqrt{-1}(\omega t + \varphi)} = \cos(\omega t + \varphi) - \sqrt{-1} \sin(\omega t + \varphi) \qquad (6.10)$$

In principle, we have a free choice between the two possible conventions, and in fact both are used. For the sake of convenience we shall adopt the convention that when the imaginary unit is used in the sense of equation (6.9) it will be denoted by i, and when it is used in the sense of equation (6.8) it will be denoted by j. We thus have the two alternative complex notations:

$$\cos(\omega t + \varphi) = e^{j(\omega t + \varphi)} \qquad (6.11)$$

and

$$\cos(\omega t + \varphi) = e^{-i(\omega t + \varphi)} \qquad (6.12)$$

If this convention were generally adopted, it would always be possible to pass from the one notation to the other by writing j for $-i$ and i for $-j$. Unfortunately, this is not possible, as the notations are not used so strictly. Each convention has its consequences; for example, if we use the notation of equation (6.11) we find that in normal low-frequency electric theory an inductance should be written $+j\omega L$, and a capacitance as $1/j\omega C$, which is the notation we are used to. If, however, the notation of equation (6.12) had been used, an inductance would become $-i\omega L$ and a capacitance $-1/i\omega C$; the meaning is the same, of course, but we are not used to seeing the expressions in this form.

The expression for a wave in the above notation often contains the factor $e^{j\omega t}$ or $e^{-i\omega t}$. In order to simplify things, this factor is often omitted, so that a travelling wave like that of equation (6.7) is simply written $E_m^1 \, e^{-j\beta x}$. In other words, a wave travelling along the x axis from left to right is written $E \, e^{-j\beta x}$ in the notation of equation (6.11), and $E \, e^{+i\beta x}$ in that of equation (6.12); this is the rather subtle reason why the i notation of equation (6.12) is often chosen in treatises on wave phenomena.

In this book, we shall adhere to the notation used in equation (6.11), and will thus denote an electric field varying sinusoidally with time by

$$E = E_0(x, y, z)\, e^{j\omega t}$$

where $E_0(x, y, z)$ is generally complex. As we have explained, this notation actually means

$$E = \mathrm{Re}\,(E_0\, e^{j\omega t})$$

6.2 Propagation of electromagnetic waves in unbounded conducting media

For our study of the propagation of electromagnetic waves through a conducting medium ($\gamma \neq 0$), we shall assume that the fields vary sinusoidally with time. For the moment, no stipulation about their variation with position will be made. The fields may thus be written

$$
\begin{aligned}
E(x, y, z, t) &= E_0(x, y, z)\, e^{j\omega t} \\
H(x, y, z, t) &= H_0(x, y, z)\, e^{j\omega t}
\end{aligned}
\tag{6.13}
$$

The terms $E_0(x, y, z)$ and $H_0(x, y, z)$ may be regarded as 'stills' of the fields E and H at time $t = 0$. They are vector quantities (characterized by a magnitude and a direction).

Here again, we shall not consider the most general case, but make the same simplifying assumptions that we made in Section 5.1: namely, that the electric field has only one non-zero component, which is a function of time and of one positional coordinate only

$$E_x = E_z = 0 \qquad \frac{\partial E_y}{\partial y} = \frac{\partial E_y}{\partial z} = 0 \tag{6.14}$$

We shall now solve Maxwell's equations for a conducting medium, subject to these assumptions.

Just as in the case $\gamma = 0$, substitution of equation (6.14) into Maxwell's equations leads to the conclusion that B_x and B_y are constant; here again we shall assume that they are both zero. Substituting these values in the solutions, we find that

$$\frac{\partial E_y}{\partial x} = - \frac{\partial B_z}{\partial t} \tag{6.15}$$

$$-\frac{\partial H_z}{\partial x} = jE_y + \frac{\partial D_y}{\partial t} \tag{6.16}$$

Differentiating equation (6.15) with respect to x and equation (6.16) with respect to t, by analogy with the previous case we find

$$\frac{\partial^2 E_y}{\partial x^2} = \mu\gamma \frac{\partial E_y}{\partial t} + \mu\varepsilon \frac{\partial^2 E_y}{\partial t^2} \tag{6.17}$$

We now introduce equation (6.13), which for the y component of the electric field has the form

$$E_y(x, y, z, t) = E_{0y}(x, y, z) \, e^{j\omega t} = E_{0y}(x) \, e^{j\omega t} \qquad (6.18)$$

Substituting this expression, and dividing both sides by $e^{j\omega t}$, we find

$$\frac{\partial^2 E_{0y}}{\partial x^2} = j\omega\mu(\gamma + j\omega\varepsilon)E_{0y} \qquad (6.19)$$

which is the wave equation for $E_{0y}(x)$ in this medium.

Let us now introduce a new variable Γ, such that

$$j\omega\mu(\gamma + j\omega\varepsilon) = \Gamma^2 \qquad (6.20)$$

Equation (6.19) can then be rewritten:

$$\frac{\partial^2 E_{0y}}{\partial x^2} = \Gamma^2 E_{0y} \qquad (6.21)$$

By a similar line of reasoning, an expression for the magnetic field can be derived from equations (6.15) and (6.16); this expression has the same form as that for the electric field. If we had not made the above simplifying assumptions, we would have found a wave equation for each of the six components of the electromagnetic field, of the form

$$\frac{\partial^2 A}{\partial x^2} + \frac{\partial^2 A}{\partial y^2} + \frac{\partial^2 A}{\partial z^2} = \Gamma^2 A$$

where $A(x, y, z)$ stands for $E_{0x}, E_{0y}, \ldots H_{0y}$ or H_{0z}.

The solution of the wave equation (6.21) is

$$E_{0y}(x) = E_m \, e^{\pm\Gamma x} \qquad (6.22)$$

where E_m is an integration constant. This can easily be verified by substitution in the wave equation.

We may thus write

$$E_y = E_{0y} \, e^{j\omega t} = E_m \, e^{j\omega t} \, e^{\pm\Gamma x} = E_m \, e^{j\omega t \pm \Gamma x} \qquad (6.23)$$

where $E = E(x, y, z, t)$ is the total electric field, of which E_y is the y component; $E_0 = E_0(x, y, z)$ is the value of E at time $t = 0$ (E_{0y} is the y component of this); and E_m is a constant, equal to the value of E_{0y} for $x = 0$.

If we substitute $\gamma = 0$, and hence $\Gamma = j\omega(\varepsilon\mu)^{\frac{1}{2}} = j\beta$, into equation (6.23) we arrive at the solution in Section 5.1, as is to be expected; the only difference is that this time it is written in the complex notation.

In general, however, $\gamma \neq 0$, and in this case it follows from the definition of Γ that

$$\Gamma = j\omega\sqrt{\varepsilon\mu(1 + \gamma/j\omega\varepsilon)} \qquad (6.24)$$

is a complex quantity, which we may represent

$$\Gamma = \alpha + j\beta \qquad (6.25)$$

The solution of the wave equation for the electric field may then be written

$$E_y = E_m e^{-\alpha x} e^{j(\omega t - \beta x)} \tag{6.26}$$

or

$$E_y = E_m e^{\alpha x} e^{j(\omega t + \beta x)} \tag{6.27}$$

This represents a travelling wave in the x_+ (or x_-) direction, as in the case where the conductivity was assumed to be zero. The amplitude $E_m e^{-\alpha x}$ (or $E_m e^{\alpha x}$) now decreases exponentially during propagation, this decrease being determined by the factor $e^{-\alpha x}$ (or $e^{\alpha x}$).

This dissipation of electromagnetic energy is naturally due to losses in the material ($\gamma \neq 0$). In a good dielectric, where the conductivity γ is low, α will also be low and the amplitude of the wave will only decrease slowly. In a fairly good conductor, where γ and hence α is large, the amplitude of the wave will decrease very rapidly. α is called the *attenuation constant* and β the *phase constant*.

It may be noted that β in a conducting medium is different from that in a perfect dielectric, even if ε and μ are the same. As γ increases, β also increases, so that the velocity of propagation decreases and the wavelength (corresponding to a given frequency) in the medium also falls.

The solution for the magnetic field can now be found from that for the electric field (equations 6.26 and 6.27), with the aid of equation (6.15). This gives

$$\frac{\partial H_z}{\partial t} = \frac{\Gamma}{\mu} E_m e^{j\omega t - \Gamma x}$$

or, after integration with respect to time

$$H_z = \frac{\Gamma}{j\omega\mu} E_m e^{j\omega t - \Gamma x} = H_{0z} e^{j\omega t} \tag{6.28}$$

The above equation represents only the magnetic field corresponding to the electric wave travelling to the right (equation 6.26). We can also write

$$H_z = \frac{\Gamma}{j\omega\mu} E_y \tag{6.29}$$

The intrinsic impedance of the medium, defined above, now becomes

$$\eta = \frac{E_y}{H_z} = \frac{j\omega\mu}{\Gamma} \tag{6.30}$$

or, with the aid of equation (6.24)

$$\eta = \sqrt{\frac{\mu}{\varepsilon(1 + \gamma/j\omega\varepsilon)}} \tag{6.31}$$

When $\gamma \neq 0$, therefore, $\eta \neq (\mu/\varepsilon)^{\frac{1}{2}}$, and in general η will be complex. Since $E_y = \eta H_z$, this means that E and H will no longer be in phase in a conducting medium, as they are in a perfect dielectric. Here η lies between $0°$ and $90°$, which means that the electric field has a phase lead with respect to the magnetic field.

The situation can of course be described in further detail by expressing α and β in terms of the material constants, ε, μ and γ and the angular frequency ω, with the aid of equations (6.24) and (6.25)

$$\alpha + j\beta = j\omega\sqrt{\varepsilon\mu(1+\gamma/j\omega\varepsilon)} \qquad (6.32)$$

Equating the real and imaginary parts of this equation, we find

$$\alpha^2 = \frac{\omega^2}{2}\,\varepsilon\mu\{-1+\sqrt{1+\gamma^2/\omega^2\varepsilon^2}\} \qquad (6.33)$$

$$\beta^2 = \frac{\omega^2}{2}\,\varepsilon\mu\{1+\sqrt{1+\gamma^2/\omega^2\varepsilon^2}\} \qquad (6.34)$$

These equations confirm clearly the behaviour of α and β as functions of γ, as described qualitatively above.

Some special cases are as follows:

1. $\gamma = 0$—then $\alpha = 0$ and $\beta = \omega(\varepsilon\mu)^{\frac{1}{2}}$, as we already found for the ideal dielectric.

2. $\gamma \ll \omega\varepsilon; \alpha \approx \frac{1}{2}\gamma(\mu/\varepsilon)^{\frac{1}{2}}$ and $\beta \approx \omega(\varepsilon\mu)^{\frac{1}{2}}$—the phase constant compared with the first case does not change greatly but attenuation is already apparent for small values of γ.

3. $\gamma \gg \omega\varepsilon$, $\alpha = \beta \approx (\omega\mu\gamma/2)^{\frac{1}{2}}$—both the phase constant and the attenuation constant increase as $\gamma^{\frac{1}{2}}$.

7 REFLECTION AND TRANSMISSION OF ELECTROMAGNETIC WAVES AT INTERFACES

Up till now we have limited ourselves to the propagation of electromagnetic waves in unbounded media, although we have derived a number of boundary conditions for the electric and magnetic fields at interfaces between two media. We shall now study the behaviour of electromagnetic waves after incidence on such interfaces by using Maxwell's equations. The main lines of this behaviour—reflection (partial or total), refraction, etc.—are known from the study of light waves.

Here again, we shall not consider the most general case of a wave polarized in an arbitrary direction and incident on the interface at an arbitrary angle, but will try to extrapolate the main ideas from a study of several simple cases.

Three such cases will be considered here:

1. A plane polarized wave incident on a perfect conductor from free space (ε_0, μ_0, $\gamma = 0$).

2. The same wave incident on a perfect dielectric from free space.

3. The same wave incident on a medium of finite conductivity from free space.

In all cases, we shall assume the media in question to be homogeneous and isotropic, the waves to be normally incident on the plane interface (so that the wave front of the plane wave—the plane of E and H—is parallel to the interface), and all fields to vary harmonically with time.

7.1 Electromagnetic waves from free space incident on ideal conductors

A plane electromagnetic wave is incident on a medium of infinite conductivity from the air. Let the air–metal interface be the plane $z = 0$ (Fig. 7.1). The electromagnetic wave is polarized in the x direction. The electric field strength of the wave is given by

$$E = E_x = E_m \, \mathrm{e}^{\mathrm{j}(\omega t - \beta z)} \tag{7.1}$$

However, we have seen above that this is only one of the solutions of the wave equation, and not the complete solution: there can also be a wave propagated along the z axis in the opposite direction. In general, therefore, the electric field above the interface can be written

$$E_x = E_m e^{j(\omega t - \beta z)} + f(\omega t + \beta z) \tag{7.2}$$

where $f(\omega t + \beta z)$ is the general expression for a wave directed upwards along the z axis (upwards in Fig. 7.1).

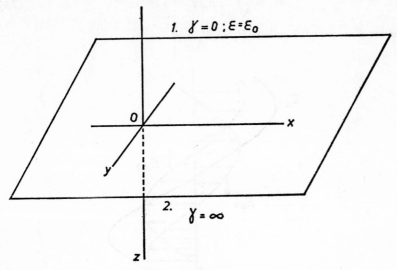

Figure 7.1

Since no field can exist in the ideal conductor ($z > 0$), and the tangential component of the electric field should be continuous at $z = 0$, we may write for $z = 0$

$$E_x(0) = 0 \tag{7.3}$$

It then follows from equation (7.2) that

$$f(\omega t) = -E_m e^{j\omega t}$$

for all arguments of f, and hence also

$$f(\omega t + \beta z) = -E_m e^{j(\omega t + \beta z)} \tag{7.4}$$

If therefore we want the boundary condition in equation (7.3) to be satisfied at the interface, the total electric field strength in the space with $z < 0$ (air) must have the form

$$E_x = E_m e^{j(\omega t - \beta z)} - E_m e^{j(\omega t + \beta z)} = E_m(e^{-j\beta z} - e^{+j\beta z}) e^{j\omega t} \tag{7.5}$$

We had already assumed that $E_y = E_z = 0$.

M.T.

3

This is shown in Fig. 7.2 for the xOz plane: the wave E_{x1} travelling downwards, the reflected wave E_{x2} and the total electric field strength E_x.

We can now determine the corresponding magnetic field with the aid of equation (7.5) and Maxwell's equations (3.18)

$$\frac{\partial H_y}{\partial t} = -\frac{1}{\mu}\frac{\partial E_x}{\partial z} = \frac{j\beta}{\mu} E_m(e^{-j\beta z}+e^{j\beta z})\,e^{j\omega t}$$

Integration with respect to time gives

$$H_y = \sqrt{\frac{\varepsilon}{\mu}} \cdot E_m(e^{-j\beta z}+e^{j\beta z})\,e^{j\omega t} \tag{7.6}$$

where $\beta = \omega\sqrt{\varepsilon\mu}$. Further, $H_x = H_z = 0$.

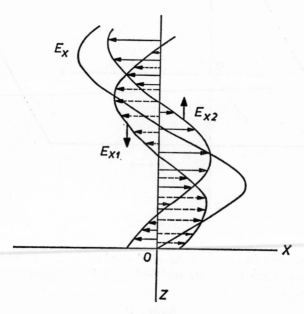

Figure 7.2

Fig. 7.3 shows the magnetic field in medium 1 (air). Unlike the electric field strength, which is reflected at the surface of the perfect conductor with a change of sign, the reflected magnetic field strength H_{y2} has the same sign as the incident H_{y1}. H_y is the resultant field strength.

The variation of E_x and H_y, the total field strengths in air, sketched in these figures represents the resultant of two travelling waves, the incident and the reflected. These two total field strengths vary continually in time (with $e^{j\omega t}$), but they do not form a travelling wave. The wave has a certain amplitude at one point and another amplitude at a new point, but for the rest it is fixed in space. Such a wave is called a *standing wave*. The electric field strength is zero at the interface $z = 0$, and points at distances of one half, one, one and a half wavelengths, etc., from the interface are called the nodes of the standing wave. Half-way between the nodes, the wave has its maximum amplitude; these points are called the anti-nodes.

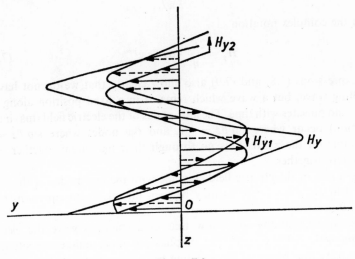

Figure 7.3

The magnetic field strength has its nodes at one quarter, three quarter, five quarter wavelengths from the interface, and so on.

Note: The existence of electromagnetic waves was experimentally determined by Hertz with the aid of this pattern of nodes and anti-nodes in a standing wave. He reflected an electromagnetic wave from a zinc sheet, and scanned the field pattern in the air with the aid of a wire frame containing a spark-gap. Sparks were produced as a result of the induced e.m.f. at the anti-nodes, but not at the nodes.

Let us now consider the form of the standing wave in somewhat greater detail.

The expression for the standing electric wave as given in equation (7.5) was

$$E_x = -E_m(e^{j\beta z} - e^{-j\beta z})\, e^{j\omega t} = -2E_m \frac{e^{j\beta z} - e^{-j\beta z}}{2j}\, j e^{j\omega t}$$

$$= -2E_m(\sin \beta z)\, je^{j\omega t} \qquad (7.7)$$

In our complex notation, this is written

$$E_x = \mathrm{Re}\,\{-2E_m\, je^{j\omega t} \sin \beta z\} = \mathrm{Re}\,\{2E_m \sin \beta z(-j\cos \omega t + \sin \omega t)\}$$

$$= 2E_m \sin \beta z \sin \omega t \qquad 7.8)$$

Similarly, we can show from equation (7.6) that the expression for the magnetic field really is

$$H_y = \frac{2E_m}{\eta} \cos \beta z \cos \omega t \qquad (7.9)$$

where

$$\eta = \left(\frac{\mu}{\varepsilon}\right)^{\frac{1}{2}}$$

or, in the complex notation

$$H_y = \frac{2E_m}{\eta} \cos \beta z \ e^{j\omega t} \qquad (7.10)$$

The expressions (7.8) and (7.9) also show clearly that we do not have a travelling wave, but a wave which remains in a fixed position along the z axis, and pulsates with time. The anti-nodes of the electric field (maximum pulsation) occur where $\sin \beta z = \pm 1$, and the nodes where $\sin \beta z = 0$. All points on the z axis thus go through their maximum together, and are at zero together.

The nodes of the electric field coincide with the anti-nodes of the magnetic field and vice versa, as may be seen directly from equations (7.8) and (7.9).

The same equations also show that in a standing wave the electric and magnetic fields are 90° out of phase in time, unlike the situation with a travelling wave. In the standing wave, the electromagnetic energy fluctuates periodically from the E field to the H field and back; the Poynting vector changes direction each quarter wavelength; there is no propagation of energy. The effect on the environment of a standing wave built up of two travelling waves is thus completely different from that of a travelling wave.

The situation found in practice with a material like copper, which while not a perfect conductor is a very good one, is quite close to that described above. However, a small part of the incident wave is not reflected, but is propagated a certain depth into the material (the penetration depth).

7.2 Electromagnetic waves from free space incident on perfect dielectrics

Let us consider a plane polarized wave incident on a perfect dielectric ($\gamma = 0$) from free space. The interface is again formed by the plane $z = 0$, as shown in Fig. 7.4. We again consider the case of normal incidence of a plane wave polarized in the x direction

$$E_{x1} = E_{m1} \ e^{j(\omega t - \beta_1 z)}$$

Part of the incident wave will normally be reflected and part will be transmitted. We shall assume that the reflected and transmitted waves have the same form as the incident wave. The incident wave is denoted by the subscript 1, the reflected wave by the subscript 2 and the transmitted wave by the subscript 3. The total electric field strength in air is thus

$$E_{xl}' = E_{m1}\, e^{j(\omega t - \beta_1 z)} + E_{m2}\, e^{j(\omega t + \beta_1 z)} \tag{7.11}$$

and in the dielectric

$$E_{xd} = E_{m3}\, e^{j(\omega t - \beta_3 z)} \tag{7.12}$$

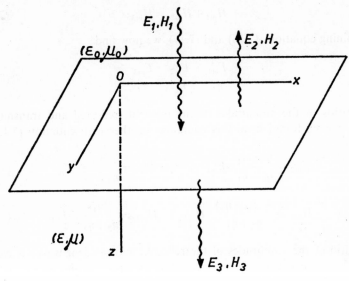

Figure 7.4

The tangential component of the electric field strength is continuous at the interface $z = 0$

$$(E_{xl})_0 = (E_{xd})_0$$

or

$$E_{m1}\, e^{j\omega t} + E_{m2}\, e^{j\omega t} = E_{m3}\, e^{j\omega t}$$

or

$$E_{m1} + E_{m2} = E_{m3} \tag{7.13}$$

The associated magnetic fields are found with the aid of Maxwell's equations

$$H_{yl} = H_{m1}\, e^{j(\omega t - \beta_1 z)} + H_{m2}\, e^{j(\omega t + \beta_1 z)} \tag{7.14}$$

$$H_{yd} = H_{m3}\, e^{j(\omega t - \beta_3 z)} \tag{7.15}$$

Once again, it follows from the equations for the three travelling waves separately that the amplitudes of E and H are coupled by the intrinsic impedance of the medium in question (see also Section 7.1)

$$E_{m1} = \eta_1 H_{m1} \qquad E_{m2} = -\eta_1 H_{m2} \qquad E_{m3} = \eta_3 H_{m3} \qquad (7.16)$$

The minus sign in the middle equation above means that E_{x2} and H_{y2} are always of opposite sign.

We shall assume further that there are no free charges in our ideal dielectric, so that the tangential magnetic fields are continuous at the interface

$$H_{m1} + H_{m2} = H_{m3} \qquad (7.17)$$

Combining equations (7.16) and (7.17), we now find

$$\frac{E_{m1}}{\eta_1} - \frac{E_{m2}}{\eta_1} = \frac{E_{m3}}{\eta_3} \qquad (7.18)$$

The ratios of the amplitudes of the incident, reflected and transmitted waves can be found from this equation together with equation (7.13)

$$E_{m2} = \frac{\eta_3 - \eta_1}{\eta_3 + \eta_1} \cdot E_{m1} \qquad E_{m3} = \frac{2\eta_3}{\eta_3 + \eta_1} \cdot E_{m1} \qquad (7.19)$$

$$H_{m2} = \frac{-(\eta_3 - \eta_1)}{\eta_3 + \eta_1} \cdot H_{m1} \qquad H_{m3} = \frac{2\eta_1}{\eta_3 + \eta_1} \cdot H_{m1} \qquad (7.20)$$

The ratio of the amplitudes of the reflected and incident waves is called the *reflection coefficient* (a term we shall encounter again) and is in this case

$$\frac{E_{m2}}{E_{m1}} = \frac{\eta_3 - \eta_1}{\eta_3 + \eta_1}$$

Numerical example: Let us take as dielectric a plastic with $\varepsilon_r \approx 4$; we then have $\varepsilon_3/\varepsilon_1 = 4$. Since $\mu_r \approx 1$ for such substances, it follows from $\eta = (\mu/\varepsilon)^{\frac{1}{2}}$ that $\eta_1/\eta_3 = \sqrt{4} = 2$. Substituting this value in the above equations, we find that the reflection coefficient is then $(\eta_3 - \eta_1)/(\eta_3 + \eta_1) = -1/3$.

These figures do not yet tell us anything about the power, as the reflection coefficient is based on amplitudes. The Poynting vector of the incident wave is in the z direction and its value per unit surface area at $z = 0$ is

$$P_1 = E_{x1} H_{y1} = E_{m1} H_{m1} \cos^2 \omega t$$

i.e. always positive or zero, with a pulsation frequency of $2f$. The maximum value of P_1 is

$$P_{m1} = E_{m1} H_{m1} = \frac{E_{m1}^2}{\eta_1}$$

Similarly, we find the maximum power of the reflected wave at $z = 0$ to be

$$P_{m2} = E_{m2} H_{m1} = \tfrac{1}{3} E_{m1} \tfrac{1}{3} H_{m1} = \frac{1}{9} \frac{E_{m1}^2}{\eta_1}$$

and that of the transmitted wave

$$P_{m3} = E_{m3} H_{m3} = \tfrac{2}{3} E_{m1} \tfrac{4}{3} H_{m1} = \frac{8}{9} \frac{E_{m1}^2}{\eta_1}$$

In other words, in our example one-ninth of the incident energy is reflected, and eight-ninths is transmitted.

In the air ($z < 0$), we now have two waves of different amplitudes travelling in opposite directions. Do these combine to give a standing wave as in Section 7.1? The answer is no, but the combination may be regarded as the superposition of a standing wave and a travelling wave.

If the reflection coefficient is nearly equal to one (large change in ε, μ or γ at the interface), the reflected wave is nearly as strong as the incident wave, and their combination is very much like a standing wave: the maximum field strength in air varies with the distance to the interface, with marked maxima and minima (see Fig. 7.5). In this connection, we may introduce an important new concept, the *voltage standing-wave ratio* (vswr), which is defined as the ratio of the amplitude of the field strength at maximum to that at minimum.

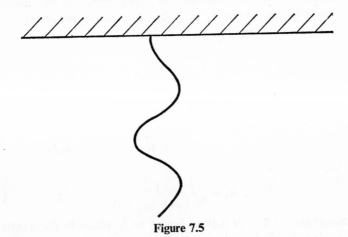

Figure 7.5

The vswr is large if the reflection coefficient is nearly equal to one. If the reflection coefficient is nearly zero (i.e. little energy reflected), the overall wave in the air above the interface looks almost like a travelling wave (Fig. 7.6). The maximum field strength varies very little from place to place in this case, and the vswr is nearly one.

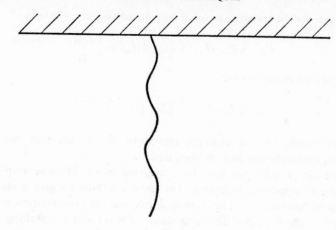

Figure 7.6

7.3 Electromagnetic waves from free space incident on media of finite conductivity

When an electromagnetic wave is incident from air on to a non-ideal conductor or a non-ideal dielectric (finite values of γ), we may adopt a line of reasoning similar to that in Section 7.2. However, we should remember that the propagation constant Γ is complex, so that the electric field strength in the conducting dielectric is written in the form

$$E_{xd} = E_{m3}\, e^{j\omega t - \Gamma_3 z} \tag{7.21}$$

and the magnetic field in the form

$$H_{yd} = H_{m3}\, e^{j\omega t - \Gamma_3 z} \tag{7.22}$$

where

$$\Gamma_3 = \alpha_3 + j\beta_3$$

In this case too we may write

$$E_{m1} = \eta_1 H_{m1} \qquad E_{m2} = -\eta_1 H_{m2} \qquad E_{m3} = \eta_3 H_{m3} \tag{7.23}$$

where η_3 is now complex

$$\eta_3 = \sqrt{\frac{\mu_3}{\varepsilon_3(1 + \gamma_3/j\omega\varepsilon_3)}} \tag{7.24}$$

This means that E and H will no longer be in phase in the transmitted wave, as a result of which the reflected wave will no longer be completely in phase with the incident wave at the interface, since the boundary conditions

$$\begin{aligned} E_{m1} + E_{m2} &= E_{m3} \\ H_{m1} + H_{m2} &= H_{m3} \end{aligned} \tag{7.25}$$

must be fulfilled.

WAVEBEARING SYSTEMS

8 GENERAL CONSIDERATIONS

We have seen so far that when an electromagnetic disturbance exists in space (for example, as the result of a point source of electromagnetic oscillations), this disturbance will be propagated in many directions. There are, however, some directions in which no radiation will be found, even if the medium is isotropic and homogeneous. The energy, which was originally concentrated in a small volume, will spread itself over an increasingly wide space. The field strength will thus become steadily less as the distance from the source increases.

If there are obstacles, such as conductors, in the way of the waves, the mode of propagation will naturally be altered. We try to make use of this fact to guide the waves in a certain direction with the aid of, say, metal structures called transmission lines (i.e. wavebearing systems). These channel the electromagnetic waves in much the same way as a megaphone channels sound waves so that they do not spread out excessively. This allows the waves to carry power or information from one place to another as high-tension cables or telegraph wires do. An unacceptably large fall in field strength at the receiving end would result from an excessive spreading.

The surface of such transmission lines acts as an interface on which the electric lines of force terminate, and in which currents can flow to provide suitable boundary conditions for the magnetic field. The wave is then propagated along the line, so that the energy (the Poynting vector) is also forced in the same direction.

Now it might be said from previous sections that since Maxwell's equations lead to plane wave-fronts propagated in a single direction, why use transmission lines? It will be remembered, however, that for the plane waves whose equations we have derived above the field strength had the same value throughout one plane perpendicular to the direction of propagation, while throughout another, parallel plane it would have another constant value. Now in theory such a plane extends to infinity, and it would therefore follow from equation (5.35) that the energy content was infinite. This is, of course, physically impossible: such plane waves represent a mathematically valid solution of the wave equation, but physically they cannot exist in free space. This nevertheless does not mean that the results derived above are valueless. We shall see that the propagation properties we have found are retained, but the field configuration becomes more complicated.

A guide or line is thus needed to force the waves into a particular direction. The transmission line must form a suitable boundary for the wave, so that it may be propagated in a limited cross-sectional area without undue distortion. In other words, the electric field must be able to end in charges on the conductor (which requires a special charge distribution on the surface of the conductor), while the magnetic field must terminate in surface currents on the conductor (see Section 4.17 for the boundary conditions which we have derived). These demands on the charge and current distribution on the conductor surface are not independent. As the wave is propagated along the conductor, the charge pattern must change continually to match the electric field pattern 'passing' at any given moment; this change in the charge distribution automatically produces surface currents, which must form suitable boundary conditions for the magnetic field of the wave.

We can illustrate this line of reasoning with a simple example. We have already derived the field pattern for a plane wave of infinite extension, propagated in the x direction in free space

$$E_y = E_m \cos{(\omega t - \beta x)}$$

$$H_z = \sqrt{\frac{\varepsilon}{\mu}} E_m \cos{(\omega t - \beta x)} \tag{8.1}$$

If we now introduce two thin conducting sheets parallel to the xOz plane (Fig. 8.1) in order to restrict the extension of the wave in one direction, the electric lines of force will end at right angles to these surfaces. We shall assume that the field, originally present for all values of y, is now restricted to the space between the sheets, and has the same value there as before.

First we can apply Maxwell's first law to the components of this field between the plates

$$-\frac{\partial H_z}{\partial x} = \frac{\partial D_y}{\partial t} \tag{8.2}$$

We have shown (equation 4.17) that the charge density σ on, for example, the lower metal surface satisfies

$$\sigma = D_y \tag{8.3}$$

If D_y now varies in time in the travelling wave according to equation (8.1), σ will also vary according to equation (8.3). If σ at a given point on the surface is to increase in this way, the extra surface charge must be provided by a surface current s_x (in A m^{-1}) to that point. This current will flow in the x direction, because D_y and hence σ vary in this direction with time. The relation between the surface charge and the current is given by the continuity equation

$$\frac{\partial s_x}{\partial x} = -\frac{\partial \sigma}{\partial t} \tag{8.4}$$

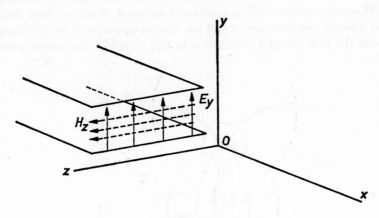

Figure 8.1

or, combining this with the previous equation

$$\frac{\partial D_y}{\partial t} = -\frac{\partial s_x}{\partial x} \tag{8.5}$$

Equation (8.5) thus determines the current required to terminate the electric field of the electromagnetic wave in the correct manner by means of the metal sheets.

Let us now consider the magnetic field. It may be seen from the boundary conditions derived in a previous section that in this special case equation (4.17) must be satisfied

$$H_t = H_z = s_x \tag{8.6}$$

where s_x has the meaning defined above.

Differentiation of equation (8.6) with respect to x gives

$$\frac{\partial H_z}{\partial x} = \frac{\partial s_x}{\partial x} \tag{8.7}$$

Combining equations (8.7) and (8.5), we see that this current will indeed provide a proper termination for the magnetic field if

$$\frac{\partial H_z}{\partial x} = -\frac{\partial D_y}{\partial t} \tag{8.8}$$

This is simply Maxwell's first law (equation 8.2), and the required condition is, therefore, always satisfied.

This means that we can limit the extension of the electromagnetic wave in one direction in the way shown in Fig. 8.1 without making the field between the metal sheets any different from that in the absence of the sheets, owing to the surface charges and currents produced in the conducting sheets. It can be proved in a similar way that transmission lines of other forms can also provide a suitable limitation of the field.

However, we have still not achieved our goal. While we have limited the wave in one direction, it still has infinite extension in the z direction. Since the field strength is constant in this direction, the energy content

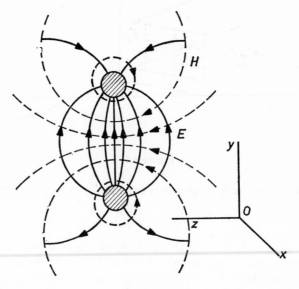

Figure 8.2

of the field will still be infinite. Anticipating a more systematic treatment, we may mention here that a different conductor configuration can be derived from that of Fig. 8.1 by rolling the top sheet upwards and the bottom sheet downwards to form two cylinders (Fig. 8.2). This gives the well-known Lecher line or parallel-wire transmission line, which can also be used as a guide for plane waves (plane waves in the sense that the only components of E and H which exist are in a plane perpendicular to the direction of propagation). The field pattern for this case is sketched for a plane parallel with the yOz plane. The electric lines of force end normally on the surface of the conductors, and form arcs of circles (with their centres on the perpendicular bisector of the line joining the centres of the two conductors). The magnetic lines of force are circles which cross all the electric lines of force at right angles.

Here too, the electromagnetic field is unlimited in extent, but the field strength falls off in all directions in such a way that the total energy content remains finite. This is thus a physically possible situation. There is one difference between this wave system and a perfectly plane wave, in that the wave must be generated at some point along the conductor, and received at another point. The wave front will certainly not be plane at these points: there will be a gradual transition from a more or less curved wave front to a plane one. Similar situations will arise at bends

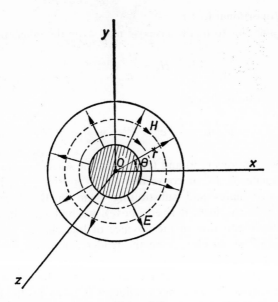

Figure 8.3

in the conductors, or sudden changes in their diameter or separation. Here the transmission line will emit a certain amount of electromagnetic energy into the surrounding space, the emission increasing with the frequency of the electromagnetic wave. Similar disturbances are produced if other objects are placed near the conductors (e.g. screening).

Another well-known transmission line is the coaxial line (we shall not consider the hollow-pipe waveguide in this qualitative discussion). In the most commonly occurring case, the electromagnetic field in the coaxial line consists of electric lines of force along the radii, and magnetic lines of force as concentric circles cross these radii at right angles (Fig. 8.3).

We shall now see how this simple field configuration can occur in this lossless case (where the field only has transverse components).

Let us consider a travelling wave in the z direction, which varies harmonically with time. The z and t dependence of the field is thus given by $e^{j(\omega t - \beta z)}$. Apart from this, E and H are only functions of r; considerations of symmetry in the simplest case indicate that the field will be independent of θ. For the inner conductor, we have

$$\oint H_1 \, dl = I$$

or

$$H \cdot 2\pi r = I$$

and H is proportional to $1/r$.

If we assume this to be the case, we may write the magnetic field

$$H = H_\theta = \frac{H_0}{r} e^{j(\omega t - \beta z)} \tag{8.9}$$

Maxwell's equations then give

$$E = E_r = \frac{E_0}{r} e^{j(\omega t - \beta z)} = \eta \frac{H_0}{r} e^{j(\omega t - \beta z)} \tag{8.10}$$

Substituting equations (8.9) and (8.10) into the wave equation written in polar coordinates (r, θ), or converting the expressions for the field into Cartesian coordinates and substituting them in the Cartesian wave equation (i.e. in the form derived above), we find that these equations satisfy the wave equation in the absence of free charge. Verification of this may be left as an exercise to the reader. Thus equations (8.9) and (8.10) form a possible field configuration. Furthermore, the boundary conditions at the inner and outer conductor must be satisfied.

The total current in the inner conductor is given by the line integral of the magnetic field strength

$$I = H_\theta \cdot 2\pi a \tag{8.11}$$

where a is the radius of the inner conductor. If the conductor is assumed to be perfect, all the current will flow on the surface, and the surface current density will be given by

$$s_z = \frac{I}{2\pi a} \, A \, m^{-1} \tag{8.12}$$

It follows from equations (8.11) and (8.12) that

$$\frac{\partial H_\theta}{\partial z} = \frac{\partial s_z}{\partial z} \tag{8.13}$$

If the surface charge density on the inner conductor is σ, the continuity equation gives

$$\frac{\partial \sigma}{\partial t} = -\frac{\partial s_z}{\partial z} \qquad (8.14)$$

And since $\sigma = D_r$, we may write

$$\frac{\partial D_r}{\partial t} = -\frac{\partial s_z}{\partial z} \qquad (8.15)$$

Combining equations (8.13) and (8.15) we find

$$\frac{\partial H_\theta}{\partial z} = -\frac{\partial D_r}{\partial t} \qquad (8.16)$$

as the condition for proper termination of the field in question on the conductor system. Now equation (8.16) follows from Maxwell's equations written in polar coordinates and this condition is, therefore, always satisfied.

The coaxial line (in its simplest form as a good conductor for a transverse field configuration, with no component of the electromagnetic field in the direction of propagation) is thus also a usable transmission line.

In the next section, we shall derive the 'telegraphers' equations' for such systems (Lecher line, coaxial line, etc.), so as to make a more quantitative treatment possible. We may remark here that this derivation is based on the assumption that the field distribution is lossless. The application of Kirchhoff's laws to equivalent circuits of these two-wire lines containing lumped self-inductances and capacitances is only permissible for lossless transmission lines, where the electromagnetic field has only transverse components. Two successive cross-sections through the line are not coupled in this case, and each line segment can thus be considered independent of the rest.

If there are losses in the line, the electric field will have an axial component, and successive line segments will no longer be independent of one another. Under these conditions, it is no longer strictly permissible to use or even to derive the telegraphers' equations. In fact, it is common to use the telegraphers' equations in such cases, with a suitable correction for the losses. It will be clear that this is only an approximation, but it will be acceptable if the losses are not excessive.

Now 'higher modes' (more complicated field configurations than the simple transverse configurations introduced above) can occur with coaxial lines as well as with waveguides. These higher modes are not dealt with by the telegraphers' equations as applied to parallel-wire lines, as the equations assume that the fields are transverse. We shall not consider higher modes at all in connection with coaxial lines; in general, they do not play a significant role. However, the reader should know how the occurrence of higher modes can be prevented. If care is taken that

$$2\pi \frac{(r_1 + r_2)}{2} < \lambda \qquad (8.17)$$

where λ is the wavelength corresponding to the operating frequency and r_1 and r_2 are the radii of the inner and outer conductor respectively, then higher modes will not occur. In other words, the mean circumference of the system should be slightly less than the wavelength.

PARALLEL-WIRE TRANSMISSION LINES

In this section, the Lecher line or the coaxial line will generally be used as examples. Much of this information is, however, applicable to other types of transverse waveguides, such as microstrip and tri-plate strip lines.

9 DERIVATION OF THE TELEGRAPHERS' EQUATIONS

On the basis of our earlier remarks on the transverse field configuration in parallel-wire systems, we may regard the transmission lines as being built up of a continuous chain of self-inductance L, capacitance C, resistance R and conductance G along the line. It should be noted that when R and G differ from zero, the argument given below is only an approximation, but a good one. We shall assume that the line is well screened, so that we do not have to consider radiation effects.

We shall call the self-inductance per unit length of the guiding system L (in henrys per metre), the capacitance of the leads with respect to one another C (in farads per metre), the loss resistance of both conductors R (in ohms per metre) and the leakage conductance, representing losses in the dielectric between the conductors, G (in siemens per metre). We shall assume that $R \neq 1/G$.

We choose the z axis along the line or cable, and divide it up by means of sections perpendicular to the z axis into short segments of length $\Delta z \ll \lambda$, where λ is the wavelength. The current i and the voltage v across the conductors in one segment will then be practically constant, while both i and v will vary with z and t down the wire.

We assume that through one cross-section the two conductors carry currents equal in magnitude but opposite in sign. The equivalent circuit for a segment Δz of the parallel-wire system will then be as sketched in Fig. 9.1.

We apply Kirchhoff's law to the branching point P, and find

$$i(z+\Delta z) = i(z) - \Delta i \tag{9.1}$$

Expanding by means of a Taylor series, we have, to a first approximation

$$i(z+\Delta z) = i(z) + \frac{\partial i}{\partial z}\,\Delta z \tag{9.2}$$

Ohm's law and the relations between the charge, current and voltage in a capacitor ($i = \mathrm{d}Q/\mathrm{d}t$ and $Q = CV$) now give

$$\Delta i = G\,\Delta z\,.\,v(z) + C\,\Delta z\,\frac{\partial v(z)}{\partial t} \tag{9.3}$$

Substituting equations (9.2) and (9.3) in equation (9.1), we find

$$-\frac{\partial i}{\partial z} = Gv + C\frac{\partial v}{\partial t} \qquad (9.4)$$

which is the *first telegraphers' equation.*

We now apply Kirchhoff's voltage law to the closed mesh $AA'B'B$, and in a similar way find

$$v(z+\Delta z) = v(z) + \frac{\partial v}{\partial z}\,\Delta z = v(z) - \Delta v = v(z) - R\,\Delta z\,i - L\,\Delta z\,\frac{\partial i}{\partial t} \quad (9.5)$$

or

$$-\frac{\partial v}{\partial z} = Ri + L\frac{\partial i}{\partial t} \qquad (9.6)$$

which is the *second telegraphers' equation.*

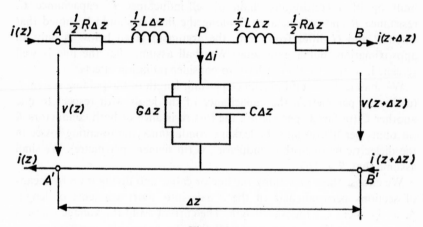

Figure 9.1

The two telegraphers' equations derived above form a set of simultaneous partial differential equations for v and i. Restricting ourselves, as above, to harmonic functions of time, and using the complex notation, we may write

$$v(z, t) = \mathrm{Re}\,\{V(z)\,e^{j\omega t}\}$$
$$i(z, t) = \mathrm{Re}\,\{I(z)\,e^{j\omega t}\} \qquad (9.7)$$

In general, the quantities V and I are complex, because they can contain an arbitrary constant phase factor. We can of course differentiate the voltage expression as follows

$$\frac{\partial v(z, t)}{\partial t} = \mathrm{Re}\,\{j\omega V(z)\,e^{j\omega t}\}$$

$$\frac{\partial v(z, t)}{\partial z} = \mathrm{Re}\,\left\{\frac{dV(z)}{dt}\,e^{j\omega t}\right\} \qquad (9.8)$$

and similarly for the current. We shall omit the Re from now on, for the sake of convenience. Substitution in the telegraph equations now gives

$$-\frac{\mathrm{d}I(z)}{\mathrm{d}z} = (G+j\omega C)V(z) \tag{9.9}$$

$$-\frac{\mathrm{d}V(z)}{\mathrm{d}z} = (R+j\omega L)I(z) \tag{9.10}$$

Our original set of partial differential equations has thus been converted into a set of ordinary differential equations.

If I and V are known for a specific point on the conductor, they can be calculated for all other points with the aid of the above equations. The calculations are facilitated if we separate V and I. For this purpose, we eliminate I by differentiating equation (9.10) with respect to z, and substitute the expression for $\mathrm{d}I/\mathrm{d}z$ found in this way into equation (9.9). This gives

$$\frac{\mathrm{d}^2V(z)}{\mathrm{d}z^2} = (R+j\omega L)(G+j\omega C)V(z) \tag{9.11}$$

Eliminating V in a similar way, we obtain

$$\frac{\mathrm{d}^2I(z)}{\mathrm{d}z^2} = (R+j\omega L)(G+j\omega C)I(z) \tag{9.12}$$

Now let us write

$$(R+j\omega L)(G+j\omega C) = \Gamma^2 = (\alpha+j\beta)^2 \tag{9.13}$$

We may now rewrite the above equations as

$$\frac{\mathrm{d}^2V(z)}{\mathrm{d}z^2} = \Gamma^2 V(z) \tag{9.14}$$

$$\frac{\mathrm{d}^2I(z)}{\mathrm{d}z^2} = \Gamma^2 I(z) \tag{9.15}$$

We have already seen above that equations of this type are wave equations, with the solutions

$$V(z) = A\,\mathrm{e}^{-\Gamma z} + B\,\mathrm{e}^{\Gamma z} \tag{9.16}$$

$$I(z) = A_1\,\mathrm{e}^{-\Gamma z} + B_1\,\mathrm{e}^{\Gamma z} \tag{9.17}$$

where A, A_1, B and B_1 are complex integration constants, which must be chosen to satisfy the boundary conditions. For example, the second telegraphers' equation and equation (9.16) give

$$I(z) = -\frac{1}{(R+j\omega L)}\frac{\mathrm{d}V}{\mathrm{d}z} = \frac{A\Gamma}{R+j\omega L}\,\mathrm{e}^{-\Gamma z} - \frac{B\Gamma}{R+j\omega L}\,\mathrm{e}^{\Gamma z} \tag{9.18}$$

Combining equations (9.17) and (9.18) now gives

$$A_1 = \frac{A\Gamma}{R+j\omega L} \quad \text{and} \quad B_1 = -\frac{B\Gamma}{R+j\omega L} \tag{9.19}$$

We now introduce a new variable, by writing

$$\frac{R+j\omega L}{\Gamma} = \sqrt{\frac{R+j\omega L}{G+j\omega C}} = Z_0 \, e^{j\psi} \tag{9.20}$$

$Z_0 \, e^{j\psi}$ is called the characteristic impedance of the transmission line; this impedance will be complex unless $R = G = 0$ or $R/L = G/C$.

We shall be returning to both the characteristic impedance and the propagation constant Γ below.

The general solution of the telegraphers' equations in our new notation is thus

$$V(z) \, e^{j\omega t} = A \, e^{-\alpha z} \, e^{j(\omega t - \beta z)} + B \, e^{\alpha z} \, e^{j(\omega t + \beta z)} \tag{9.21}$$

$$I(z) \, e^{j\omega t} = \frac{A}{Z_0 \, e^{j\psi}} \, e^{-\alpha z} \, e^{j(\omega t - \beta z)} - \frac{B}{Z_0 \, e^{j\psi}} \, e^{\alpha z} \, e^{j(\omega t + \beta z)} \tag{9.22}$$

Here Γ has been written as $(\alpha + j\beta)$, and both sides of the equations have been multiplied by $e^{j\omega t}$.

Both the current and the voltage along the transmission line considered here can in general be written as a superposition of two travelling waves in the z_+ and z_- directions; these waves will generally be damped.

The significance of the different variables introduced above can be made clearer by illustration (Fig. 9.2). Let us consider, say, the voltage wave travelling to the right

$$A . e^{-\alpha z} \, e^{j(\omega t - \beta z)}$$

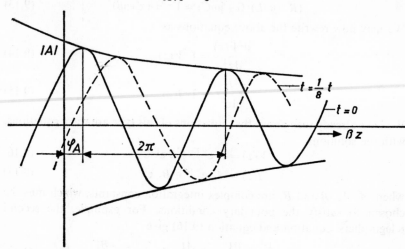

Figure 9.2

The amplitude of this wave decreases exponentially in the z_+ direction. The planes of constant phase are given by

$$\omega t - \beta z = \text{constant}$$

They thus travel in the z_+ direction with phase velocity

$$v_{\text{ph}} = \frac{dz}{dt} = \frac{\omega}{\beta} \tag{9.23}$$

The voltage of this wave at $t = 0$ will be

$$v(z, 0) = \text{Re} \{A\, e^{-\alpha z}\, e^{-j\beta z}\}$$

Writing $A = |A|\cdot e^{j\phi_A}$, we find

$$v(z, 0) = |A|\, e^{-\alpha z} \cos(\varphi_A - \beta z) \tag{9.24}$$

This relationship is plotted as the continuous line in Fig. 9.2. The broken line represents the same wave at a slightly later time, $t = T/8 = 1/8f = 2\pi/8\omega = \pi/4\omega$. The voltage of this wave is then

$$v\left(z, \frac{\pi}{4\omega}\right) = v\left(z, \frac{T}{8}\right) = |A|\, e^{-\alpha z} \cos(\varphi_A + \pi/4 - \beta z)$$

Fig. 9.2 gives the real voltage $v(z, t)$. It is also possible to plot the complex voltage $V(z)$ as a function of z in the complex plane; this gives a spiral (Fig. 9.3). The modulus and argument of V decrease as z increases

$$V(z) = A\, e^{-\alpha z}\, e^{-j\beta z} = |A|\, e^{-\alpha z}\, e^{j\varphi_A}\, e^{-j\beta z} \tag{9.25}$$

If $R = G = 0$, α will be zero too, and the spiral then changes into a circle. This means that the amplitude of the wave remains constant as it travels down the line.

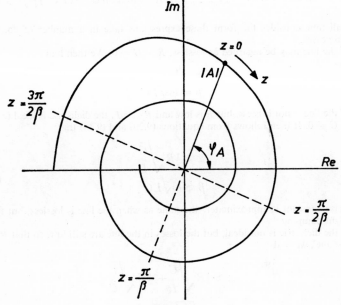

Figure 9.3

9.1 The propagation constant

The quantity $\Gamma = \alpha + j\beta$ defined in equation (9.13) is called the propagation constant. As we have seen above, the real part α determines the damping as waves pass along the line, α being the damping constant. The phase of the waves at a given point on the transmission line is determined by the imaginary part β, among other factors; β is the phase constant.

We shall now express α and β in terms of the line parameters L, C, R and G, and the frequency, with the aid of equation (9.13). Separating this equation into its real and imaginary parts, we obtain

$$\alpha^2 - \beta^2 = RG - \omega^2 LC \tag{9.26}$$

$$2\alpha\beta = \omega(RC + LG) \tag{9.27}$$

The right-hand side of equation (9.27) is positive, which means that α and β must both be of the same sign. In this case, we choose the positive sign, since the amplitude of the waves would otherwise increase along the transmission line, instead of being damped.

Squaring equations (9.26) and (9.27) and adding, we find

$$\alpha^2 + \beta^2 = [(R^2 + \omega^2 L^2)(G^2 + \omega^2 C^2)]^{\frac{1}{2}} \tag{9.28}$$

Combining equations (9.26) and (9.28), we see that in general

$$\alpha = \{\tfrac{1}{2}(RG - \omega^2 LC) + \tfrac{1}{2}[(R^2 + \omega^2 L^2)(G^2 + \omega^2 C^2)]^{\frac{1}{2}}\}^{\frac{1}{2}} \tag{9.29}$$

$$\beta = \{-\tfrac{1}{2}(RG - \omega^2 LC) + \tfrac{1}{2}[(R^2 + \omega^2 L^2)(G^2 + \omega^2 C^2)]^{\frac{1}{2}}\}^{\frac{1}{2}} \tag{9.30}$$

We shall now consider the form these expressions take in a number of commonly occurring cases.

1. If the line may be regarded as lossless, $R = G = 0$. We then find

$$\alpha = 0$$
$$\beta = \omega\sqrt{LC} \tag{9.31}$$

2. If the line is not lossless, but R is low and $R \ll \omega L$, the dielectric is ideal (e.g. air), so that $G = 0$. It then follows from equations (9.29) and (9.30) that

$$\alpha \approx \tfrac{1}{2}R\sqrt{\frac{C}{L}}$$
$$\beta \approx \omega\sqrt{LC} \tag{9.32}$$

The phase constant is approximately the same as when the line is lossless, but there is now slight damping.

3. If the dielectric is not ideal, but the losses in the line are still low, so that $R \ll \omega L$ and $G \ll \omega C$, we find

$$\alpha \approx \tfrac{1}{2}R\sqrt{\frac{C}{L}} + \tfrac{1}{2}G\sqrt{\frac{L}{C}}$$
$$\beta \approx \omega\sqrt{LC} \tag{9.33}$$

In fact, the complete expression for β is

$$\beta = \omega\sqrt{LC}\left\{1 + \left(\frac{R}{2\omega L} - \frac{G}{2\omega C}\right)^2\right\}^{\frac{1}{2}} \tag{9.34}$$

The phase constant still has the same value, to a first approximation, but the damping is slightly greater than in the previous case.

4. If the line parameters satisfy the special relation

$$\frac{R}{L} = \frac{G}{C} \tag{9.35}$$

we have exactly

$$\alpha = R\sqrt{\frac{C}{L}}$$

$$\beta = \omega\sqrt{LC} \tag{9.36}$$

9.2 Phase velocity and group velocity

We saw above (equation 9.23) that the phase velocity is given by $v_{ph} = \omega/\beta$.

Substituting the expression (9.34) for β, which applies where the losses in transmission line and dielectric are low, we find

$$v_{ph} \approx \frac{1}{\sqrt{LC}} \cdot \frac{1}{1 + \frac{1}{8}\left(\frac{R}{\omega L} - \frac{G}{\omega C}\right)^2} \tag{9.37}$$

In the lossless case, where equation (9.35) is valid, the phase velocity becomes

$$v_{ph} = 1/\sqrt{LC} \tag{9.38}$$

In the lossless case (and in the special case of equation 9.35), therefore, the phase velocity is independent of the frequency. The line is then said to be free of *dispersion*. In the general case where losses are present, however, v_{ph} depends on the frequency, as we have seen in equation (9.37), and the line then exhibits dispersion. If a finite frequency band, not just a single frequency, is propagated along such a dispersive line, the wave will be not only attenuated but also distorted: not all the frequencies in the band will arrive at the output at the same time. In order to describe what happens when a finite frequency band is propagated along a transmission line, we need another velocity—the *group velocity*. We shall now explain this concept with reference to the simple case of a wave with only two singular harmonic components propagated in the z_+

direction. (In general, of course, a finite frequency band contains many more than two components.) We assume further that the two components have the same amplitude A but slightly different frequencies

$$\omega_1 = \omega_0 + \Delta\omega$$

and
$$\omega_2 = \omega_0 - \Delta\omega \tag{9.39}$$

where $\Delta\omega \ll \omega_0$.

We assume the damping constant α to be zero, and the phase constants for the two components to be

$$\beta_1 = \beta_0 + \Delta\beta$$

and
$$\beta_2 = \beta_0 - \Delta\beta \tag{9.40}$$

where $\Delta\beta \ll \beta_0$.

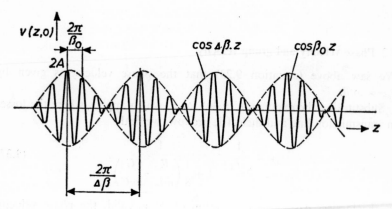

Figure 9.4

The two components of the wave may thus be written

$$V_1 = A\,e^{j(\omega_1 t - \beta_1 z)}$$
$$V_2 = A\,e^{j(\omega_2 t - \beta_2 z)} \tag{9.41}$$

Let us now assume for the sake of convenience that the two waves are in phase at $t = 0$. The total wave in the z_+ direction is thus

$$V = V_1 + V_2 = A(e^{j(\omega_1 t - \beta_1 z)} + e^{j(\omega_2 t - \beta_2 z)})$$
$$= A\,e^{j(\omega_0 t - \beta_0 z)}\{e^{j(t\Delta\omega - z\Delta\beta)} + e^{-j(\Delta\omega t - \Delta\beta z)}\}$$
$$= 2A\{\cos(t\Delta\omega - z\Delta\beta)\}e^{j(\omega_0 t - \beta_0 z)} \tag{9.42}$$

This equation represents a travelling wave of high frequency ω_0 in the z_+ direction, with phase velocity $v_{\text{ph}} = \omega_0/\beta_0$, while the envelope of the wave is itself a wave, with the lower frequency $\Delta\omega$ (see also Fig. 9.4).

The velocity of this envelope is called the group velocity v_g

$$v_g = \frac{dz}{dt} = \frac{\Delta\omega}{\Delta\beta} = \frac{1}{\Delta\beta/\Delta\omega} \approx \frac{1}{d\beta/d\omega} \qquad (9.43)$$

If the losses in the transmission line are zero, so that $\beta = \omega\sqrt{LC}$, the phase and group velocities are equal

$$v_{ph} = \frac{\omega}{\beta} = \frac{1}{\sqrt{LC}} \qquad (9.44)$$

$$v_g = \frac{1}{d\beta/d\omega} = \frac{1}{\sqrt{LC}} \qquad (9.45)$$

In general, however, v_{ph} and v_g are not equal, and the group velocity is lower than the phase velocity.

The following example may help to clarify the above. Imagine you are standing at the sea-shore on a long stretch of straight beach. The wind is blowing towards you, nearly at right angles to the shore, and the waves are coming slowly towards you, their front being almost parallel to the shore. The velocity with which these waves advance is the 'group velocity'. However, if you look along the beach and see a wave of much higher velocity, the point at which this wave meets the shore moves along the beach very quickly. The rate of this movement is the phase velocity.

The group velocity is a more material, physical quantity, always less than the velocity of light, while the phase velocity may be greater than that of light.

10 GENERAL PROPERTIES OF PARALLEL-WIRE TRANSMISSION LINES

10.1 Infinitely long transmission lines; characteristic impedance

Let us consider a transmission line extending infinitely in the z_+ direction, with a voltage source V_0 at the input, $z = 0$ (Fig. 10.1).

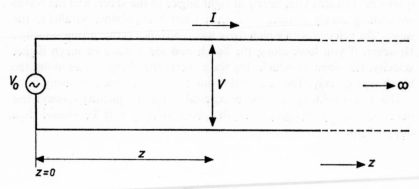

Figure 10.1

It is an easy matter to determine the integration constants A and B of equations (9.21) and (9.22) for the boundary conditions which apply to this situation.

If B were not zero, the amplitude of V and I would increase with increasing z, i.e. with increasing distance from the source V_0 at $z = 0$. This is not physically possible, so we must conclude that $B = 0$.

The voltage and the current along the transmission line are then given by

$$V = A\,e^{-\alpha z}\,e^{-j\beta z} = A\,e^{-\Gamma z} \tag{10.1}$$

$$I = \frac{A}{Z_0\,e^{j\psi}}\,e^{-\alpha z}\,e^{-j\beta z} = \frac{A}{Z_0\,e^{j\psi}}\,e^{-\Gamma z} \tag{10.2}$$

Now $V = V_0$ at $z = 0$. It follows that

$$V_0 = A \tag{10.3}$$

and hence

$$V = V_0 \, e^{-\Gamma z}$$
$$I = \frac{V_0}{Z_0 \, e^{j\psi}} \, e^{-\Gamma z} \tag{10.4}$$

We see from this that the input impedance of the line in the z_+ direction at a point z along the line is given by

$$Z_i = \frac{V(z)}{I(z)} = Z_0 \, e^{j\psi} \tag{10.5}$$

The input impedance of an infinitely long homogeneous transmission line is thus equal to the characteristic impedance of the line. The instantaneous values of the current and voltage may be found from equation (10.4) in the usual way

$$V = V_0 \, e^{-\alpha z} \cos(\omega t - \beta z)$$
$$I = \frac{V_0}{Z_0} \, e^{-\alpha z} \cos(\omega t - \beta z - \psi) \tag{10.6}$$

where Z_0 is the modulus and ψ the argument of the characteristic impedance (which will in general be complex when there are losses in the line). A further important conclusion may be drawn from the above line of reasoning: if a transmission line of finite length is terminated with its characteristic impedance, it will behave as if it were of infinite length, and there will only be one travelling wave (Fig. 10.2). The form of the wave along such a transmission line at a given moment is sketched in Fig. 10.3.

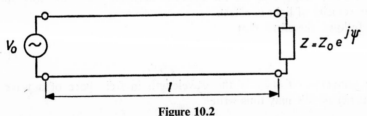

Figure 10.2

By definition, the wavelength is the distance between two corresponding points on the wave, e.g. between two successive points at which the voltage goes through zero in the same direction (Fig. 10.3).

It then follows from equation (10.6) that

$$\beta\lambda = 2\pi \tag{10.7}$$

In most cases, certainly when there are no losses, we may write

$$\beta \approx \omega\sqrt{LC}$$

Now

$$\lambda = 2\pi/\omega\sqrt{LC}$$

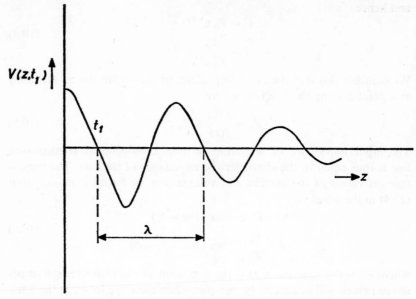

Figure 10.3

We shall show below that

$$LC = \varepsilon\mu = \varepsilon_r \mu_r / c^2 \tag{10}$$

for this transmission line, where ε and μ are the dielectric constant and permeability respectively of the material between the conductors and c is the velocity of light *in vacuo*.

It follows, therefore, that

$$\lambda = \frac{2\pi c}{\omega} \frac{1}{\sqrt{\varepsilon_r \mu_r}} \tag{10.9}$$

Now $2\pi c/\omega = c/f = \lambda_0$ is the wavelength in free space of a wave of frequency ω. We may thus write

$$\lambda = \lambda_0/\sqrt{\varepsilon_r \mu_r} \tag{10.10}$$

In a transmission line with a medium of ε_r and μ_r between the conductors, the wavelength is a factor $(\varepsilon_r \mu_r)^{\frac{1}{2}}$ smaller than in free space. If the medium is air, the wavelength will be the same as in free space.

Finally, a few words about the characteristic impedance, defined in equation (9.20).

If the losses are zero ($R = G = 0$), or if $R/L = G/C$, the characteristic impedance is real and equal to

$$Z_0 = \sqrt{\frac{L}{C}} \tag{10.11}$$

If, as is often the case, G may be put equal to 0, but $R \neq 0$ and $R \ll \omega L$, then it follows from equation (9.20) that

$$Z_0 \approx \sqrt{\frac{L}{C}}$$

$$\psi = -\arctan \frac{R}{2\omega L} \approx -\frac{R}{2\omega L} \tag{10.12}$$

In other words, when the losses are slight the characteristic impedance becomes slightly complex, but the imaginary part is so small that it is often negligible. The magnitude of the modulus does not change in the first instance.

10.2 Calculation of line constants L and C

Firstly we consider the Lecher line, which consists of two parallel conductors (of diameter $2r$) a distance a apart, as shown in Fig. 10.4. We shall assume that $a \gg r$, in other words that the conductors are so far apart in comparison with their diameter that the charge distribution on one is not influenced by that on the other.

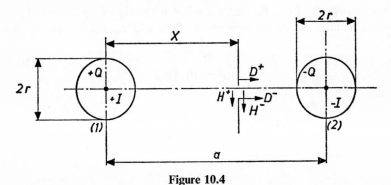

Figure 10.4

This means that, for example, at the point x on the line joining the centres of the two conductors the total dielectric displacement D is composed of two independent contributions D^+ and D^-, from the conductors (1) and (2) respectively. Similarly, the charge distribution over the surface of the conductors is uniform. D^+ is directed away from the conductor of charge $+Q$, while D^- is directed towards that of charge $-Q$.

If we apply equation (3.5) to a cylindrical surface round conductor (1), cutting the line joining the centres at x, of unit length in the longitudinal direction and concentric with the conductor, we find

$$D^+ 2\pi x = +Q \tag{10.13}$$

For a similar cylinder round conductor (2) we have

$$D^- 2\pi x = +Q \tag{10.14}$$

This means that

$$D_x = D^+ + D^- = \frac{Q}{2\pi}\left(\frac{1}{x} + \frac{1}{a-x}\right) \tag{10.15}$$

The voltage V between the conductors at the point x is found from

$$V = \int_r^{a-r} E \, dx = \int_r^{a-r} \frac{D}{\varepsilon} \, dx = \frac{Q}{2\pi\varepsilon} \int_r^{a-r} \left(\frac{1}{x} + \frac{1}{a-x}\right) dx$$

$$= \frac{Q}{2\pi\varepsilon}\left[\ln x - \ln(a-x)\right]_r^{a-r} = \frac{Q}{\pi\varepsilon} \ln \frac{a-r}{r} \tag{10.16}$$

The capacitance per unit length is of course given by

$$C = \frac{Q}{V} = \frac{\pi\varepsilon}{\ln(a-r)/r} \tag{10.17}$$

Furthermore, the self-inductance per unit length by definition is

$$L = \Phi/I \tag{10.18}$$

The magnetic field can be determined with the aid of equation (3.1) in a similar manner to the determination of the electric field as given above

$$2\pi x H^+ = I$$
$$2\pi(a-x)H^- = I \tag{10.19}$$

Hence

$$H = H^+ + H^- = \frac{I}{2\pi}\left(\frac{1}{x} + \frac{1}{a-x}\right) \tag{10.20}$$

$\Phi = \int B_n \, dS$ is thus given by

$$\Phi = \int_r^{a-r} B \, dx = \mu \int_r^{a-r} H \, dx = \frac{\mu I}{2\pi} \int_r^{a-r} \left(\frac{1}{x} + \frac{1}{a-x}\right) dx = \frac{\mu I}{\pi} \ln \frac{a-r}{r} \tag{10.21}$$

which, together with equation (10.18), gives

$$L = \frac{\mu}{\pi} \ln \frac{a-r}{r} \tag{10.22}$$

The product LC, which occurs so frequently, is then given by equations (10.17) and (10.22)

$$LC = \varepsilon\mu$$

We shall now determine the line constants for the coaxial cable (Fig. 10.5) in a similar way. The electric lines of force go from a positive charge to

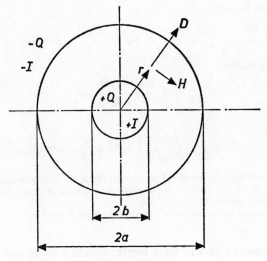

Figure 10.5

a negative one. We find

$$2\pi D = Q \tag{10.23}$$

$$V = \int_b^a E \, dr = \int_b^a \frac{Q}{2\pi\varepsilon} \frac{dr}{r} = \frac{Q}{2\pi\varepsilon} \ln \frac{a}{b} \tag{10.24}$$

$$C = \frac{Q}{V} = \frac{2\pi\varepsilon}{\ln (a/b)} \tag{10.25}$$

And for the magnetic field

$$2\pi r H = I \tag{10.26}$$

$$\Phi = \int_b^a B \, dr = \int_b^a \frac{\mu I}{2\pi} \frac{dr}{r} = \frac{\mu I}{2\pi} \ln \frac{a}{b} \tag{10.27}$$

$$L = \frac{\Phi}{I} = \frac{\mu}{2\pi} \ln \frac{a}{b} \tag{10.28}$$

Here, too, the product of L and C is equal to $\varepsilon\mu$. In air transmission, of course, this product is $\varepsilon_0\mu_0 = 1/c^2$.

The phase velocity in lossless lines is thus

$$v_{ph} = 1/\sqrt{LC} = 1/\sqrt{\varepsilon\mu} = c/\sqrt{\varepsilon_r\mu_r} \tag{10.29}$$

The characteristic impedance of a lossless line is equal to $Z_0 = \sqrt{L/C}$, which can also be expressed in terms of the dimensions of the line and the properties of the dielectric. For Lecher lines

$$Z_0 = \frac{1}{\pi} \sqrt{\frac{\mu}{\varepsilon}} \ln \frac{a-r}{r} \tag{10.30}$$

which in air becomes

$$Z_0 = \frac{1}{\pi} \sqrt{\frac{\mu_0}{\varepsilon_0}} \ln \frac{a-r}{r} = 120 \ln \frac{a-r}{r} \approx 276 \log_{10} \left(\frac{a-r}{r} \right) \quad (10.31)$$

For coaxial cables

$$Z_0 = \frac{1}{2\pi} \sqrt{\frac{\mu}{\varepsilon}} \ln \frac{a}{b} \quad (10.32)$$

and in air

$$Z_0 = \frac{1}{2\pi} \sqrt{\frac{\mu_0}{\varepsilon_0}} \ln \frac{a}{b} = 60 \ln \frac{a}{b} \approx 138 \log_{10} \frac{a}{b} \quad (10.33)$$

Note: It is difficult for technical reasons to make the characteristic impedance of a coaxial cable much less than about 10 ohms or much more than a few hundred ohms. Common values are 50, 75 and 135 ohms.

It is often easier to realize lower impedance values with strip lines.

10.3 Transmission lines of finite length; standing waves and reflection

Note: Unless otherwise stated, we shall here assume that the transmission lines are lossless, in order to avoid complicating the calculations needlessly. In practice, transmission lines are generally lossless to a first approximation, and the presence of slight losses should not change the situation appreciably.

Let us consider a line of length l, supplied by a voltage source V_0 at $z = 0$ and terminated by an impedance Z_l at $z = l$. The voltage and current at $z = l$ will be denoted by V_l and I_l respectively (see Fig. 10.6).

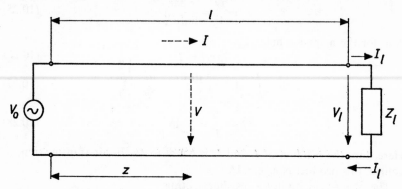

Figure 10.6

We have already found the general expressions for the voltage and current along a transmission line of characteristic impedance Z_0

$$V(z) = V^+(z) + V^-(z) = A\,e^{-j\beta z} + B\,e^{j\beta z} \quad (10.34)$$

$$I(z) = I^+(z) + I^-(z) = \frac{A}{Z_0}\,e^{-j\beta z} - \frac{B}{Z_0}\,e^{j\beta z} \quad (10.35)$$

Unlike the infinitely long transmission line, or the line terminated by its characteristic impedance, the transmission line considered here has $B \neq 0$, and travelling voltage and current waves are possible in both directions.

We shall now determine the integration constants for the boundary conditions prevailing at the two ends of the transmission line under consideration. For $z = l$ we have

$$V(l) = A\,e^{-j\beta l} + B\,e^{j\beta l} \tag{10.36}$$

$$I(l) = \frac{A}{Z_0}\,e^{-j\beta l} - \frac{B}{Z_0}\,e^{j\beta l} \tag{10.37}$$

whence we see without difficulty

$$B = A\,\frac{Z_l - Z_0}{Z_l + Z_0}\,e^{-j2\beta l} \tag{10.38}$$

where

$$Z_l = \frac{V(l)}{I(l)} \tag{10.39}$$

Further, at $z = 0$

$$V_0 = A + B \tag{10.40}$$

so that

$$A = \frac{V_0}{1 + \dfrac{Z_l - Z_0}{Z_l + Z_0} \cdot e^{-2j\beta l}} \tag{10.41}$$

Let us now call

$$r = \frac{B\,e^{j\beta l}}{A\,e^{-j\beta l}} = \frac{Z_l - Z_0}{Z_l + Z_0} \tag{10.42}$$

the *reflection coefficient* (see also Section 7.2)—r is in general a complex quantity.

Instead of equations (10.34) and (10.35) we can write

$$V(z) = V_0\,\frac{Z_l \cos \beta(l-z) + jZ_0 \sin \beta(l-z)}{Z_l \cos \beta l + jZ_0 \sin \beta l} \tag{10.43}$$

$$I(z) = \frac{V_0}{Z_0}\,\frac{Z_0 \cos \beta(l-z) + jZ_l \sin \beta(l-z)}{Z_l \cos \beta l + jZ_0 \sin \beta l} \tag{10.44}$$

and with the aid of these equations we can determine the voltage and the current at a given point z along the line in terms of the boundary conditions and line constants of this finite line.

For example, the voltage can also be written as follows

$$V(z) = V^{+}(z) + V^{-}(z) = A\,e^{-j\beta z} + Ar\,e^{j\beta(z-2l)} \tag{10.45}$$

This represents the superposition of two travelling waves V^+ and V^-

$$V^+(z) = A\,e^{-j\beta z}$$
$$V^-(z) = Ar\,e^{j\beta(z-2l)} \tag{10.46}$$

$V^+(z)$ is the incident or direct wave, and $V^-(z)$ the reflected wave. If, as we have assumed, there are no losses, the amplitudes of both waves along the line are constant. The instantaneous values can be written

$$v^+(z, t) = \operatorname{Re} V^+(z)\,e^{j\omega t} = A\cos(\omega t - \beta z)$$
$$v^-(z, t) = \operatorname{Re} V^-(z)\,e^{j\omega t} = A|r|\cos(\omega t + \beta z - 2\beta l + \varphi_r) \tag{10.47}$$

where we assume for the moment that A is real, for the sake of convenience, and the reflection coefficient is denoted by $r = |r|.e^{j\phi_r}$.

We find from equation (10.46) for the end of the line $z = l$ that

$$V^+(l) = A\,e^{-j\beta l}$$
$$V^-(l) = rA\,e^{-j\beta l} \tag{10.48}$$

whence the significance of the reflection coefficient again appears

$$r = \frac{V^-(l)}{V^+(l)} = \frac{Z_l - Z_0}{Z_l + Z_0} \tag{10.49}$$

The reflection coefficient is the ratio of the complex amplitudes of the reflected and incident waves at the closed end of the line. This coefficient is thus a measure of the extent to which the incident wave is reflected; both the amplitude and phase of the reflection are represented in the coefficient.

Some simple examples are as follows:

1. If the line is terminated by its characteristic impedance, $Z_l = Z_0$, then $r = 0$. There is no reflection, only a travelling wave in the z_+ direction.

2. If the line is open at $z = l$, so that $Z_l \to \infty$, then $r = +1$. The incident and reflected waves have the same amplitude and phase.

3. If the line is short-circuited at $z = l$, so that $Z_l = 0$, then $r = -1$. The incident and reflected waves are in anti-phase, and have the same amplitude.

4. If the line is terminated with a reactance at $z = l$, so that $Z_l = jX_l$, then $|r| = 1$, while the phase of r depends on the value of X_l. The reflected wave has the same amplitude as the incident, but a different phase.

Alternatively, it is easy to express Z_l in terms of Z_0 and r, with the aid of equation (10.49)

$$Z_l = Z_0\,\frac{1+r}{1-r} \tag{10.50}$$

Generalizing, we find for an arbitrary point on the line

$$Z(z) = Z_0 \frac{1+r(z)}{1-r(z)} \qquad (10.51)$$

or

$$r(z) = \frac{Z(z)-Z_0}{Z(z)+Z_0} \qquad (10.52)$$

Moreover, equation (10.49) also makes it easy to express the reflection coefficient at the point z in terms of the value of r at the closed end of the line ($z = l$). We have

$$r(l) = \frac{V^-(l)}{V^+(l)} = \frac{B\,e^{j\beta l}}{A\,e^{-j\beta l}}$$

and hence

$$r(z) = \frac{B\,e^{j\beta z}}{A\,e^{-j\beta z}} = r(l).e^{-2j\beta(l-z)} \qquad (10.53)$$

So far, we have been considering the total voltage or current wave as being separated into two travelling waves. Of course, these are really combined to give a total wave with maxima and minima at discrete distances from one another (see also Chapter 7), and standing waves are produced.

In order to determine the form of the total wave, we have to find the resultant of the complex amplitudes of the incident and reflected waves

$$V(z) = V^+(z) + V^-(z)$$

Now we plot

$$V^+(z) = |A|\,e^{-j\beta z + j\varphi_A}$$

and

$$V^-(z) = |A||r|\,e^{j\beta z - j2\beta l + \varphi_A + \varphi_r} \qquad (10.54)$$

in the complex plane (Fig. 10.7).

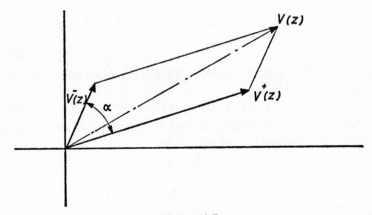

Figure 10.7

The angle α between the two vectors V^+ and V^- is

$$\alpha = (\beta z - 2\beta l + \varphi_A + \varphi_r) - (-\beta z + \varphi_A)$$
$$= 2\beta z - 2\beta l + \varphi_r \qquad\qquad 10.55)$$

where $$A = |A| \cdot e^{j\Phi_A} \qquad\qquad (10.56)$$

By using the cosine rule, we find

$$|V(z)|^2 = |V^-|^2 + |V^+|^2 + 2|V^-||V^+| \cos \alpha \qquad (10.57)$$

or

$$|V(z)| = |A| \sqrt{1 + |r|^2 + 2|r| \, \cos (2\beta z - 2\beta l + \varphi_r)} \qquad (10.58)$$

In Fig. 10.8, the modulus (amplitude) of the total complex voltage is

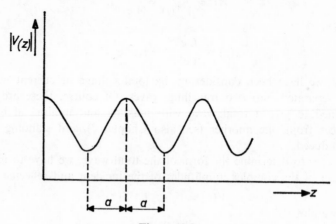

Figure 10.8

plotted as a function of position along the transmission line. At a given point on the line, of course, the instantaneous value of the voltage varies in time with an angular frequency ω.

It may be seen from equation (10.58) that $|V(z)|$ is maximum when $\cos \alpha = 1$, i.e. when

$$2\beta z - 2\beta l + \varphi_r = 2k\pi \qquad\qquad (10.59)$$

where $k = 0, \pm 1, \pm 2$, etc., and that $|V(z)|$ is minimum when $\cos \alpha = -1$, i.e. when

$$2\beta z - 2\beta l + \varphi_r = (2k+1)\pi \qquad\qquad (10.60)$$

The distance a between a maximum and a minimum in the standing wave follows from equations (10.59) and (10.60)

$$2\beta a = \pi$$

or

$$a = \pi/2\beta = (\pi/2)(\lambda/2\pi) = \lambda/4 \qquad\qquad (10.61)$$

where λ is the wavelength of the waves in question.

The magnitudes of the maxima and minima in the standing wave depend on the size of $|r|$.

For the maximum, equations (10.58) and (10.59) give

$$|V(z)|_{max} = |A|\{1+|r|\} \tag{10.62}$$

and for the minimum, equations (10.58) and (10.60) give

$$|V(z)|_{min} = |A|\{1-|r|\} \tag{10.63}$$

A quantity which is often encountered, and which can be measured quite reliably, is the *voltage standing-wave ratio* η, the ratio of the modulus of the complex voltage at the maximum to that at the minimum

$$\eta = \frac{|V|_{max}}{|V|_{min}} = \frac{1+|r|}{1-|r|} \tag{10.64}$$

Since $|r| \leqslant 1$, it follows that $\eta \geqslant 1$.

If, as happens in certain cases, the line is terminated by a real impedance $Z_l = R_l$, it follows from equations (10.49) and (10.64) that the voltage standing-wave ratio is

$$\eta = R_l/Z_0 \quad \text{or} \quad Z_0/R_l \tag{10.65}$$

depending on whether $R_l > Z_0$ or $R_l < Z_0$.

The reflection coefficient r completely determines the voltage standing-wave ratio, but conversely η only determines the value of $|r|$. The voltage standing-wave ratio is determined by scanning the field with the aid of a probe, thus finding the ratio of the field strength at a maximum to that at a minimum.

If the transmission line is terminated with a reactance, i.e. $Z_l = jX_l$, then $|r| = 1$, so $\eta \to \infty$. The amplitudes of the waves in the two directions are equal, and hence the minima in the standing wave are zero.

In the above, we have considered only voltages. It will be clear that the same arguments can be applied to the current waves. For example, the expression for the modulus of the complex total current is similar to that for the voltage given in equation (10.58)

$$|I(z)| = \frac{|A|}{Z_0} \sqrt{1+|r|^2-2|r| \cos(2\beta z - 2\beta l + \varphi_r)} \tag{10.66}$$

The current maxima occur together with the voltage minima and vice versa.

10.4 Input impedance

In this section we shall derive an expression for the input impedance Z_i of a line of length l, terminated with an impedance Z_l (see Fig. 10.9). The characteristic impedance of the line is Z_0. We gave an expression for the current through the conductor at the point z in equation (10.44). It follows from this that when $z = 0$, the current is

$$I(0) = \frac{V_0}{Z_0} \frac{Z_0 \cos \beta l + jZ_l \sin \beta l}{Z_l \cos \beta l + jZ_0 \sin \beta l}$$ (10.67)

where V_0 is the voltage at $z = 0$.

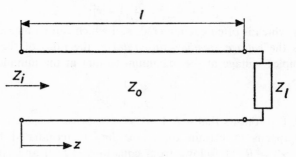

Figure 10.9

The input impedance at $z = 0$ is naturally equal to the ratio of the complex voltage and current at $z = 0$

$$Z_i = \frac{V(0)}{I(0)} = \frac{V_0}{I(0)} = Z_0 \frac{Z_l \cos \beta l + jZ_0 \sin \beta l}{Z_0 \cos \beta l + jZ_l \sin \beta l}$$

or

$$Z_i = Z_0 \frac{Z_l + jZ_0 \tan \beta l}{Z_0 + jZ_l \tan \beta l}$$ (10.68)

If the transmission line had not been lossless, we would have found

$$Z_i = \frac{Z_l + Z_0 \tanh \Gamma l}{Z_0 + Z_l \tanh \Gamma l}$$ (10.69)

A few simple examples, of very common occurrence:

1. The line is terminated at $z = l$ with its characteristic impedance Z_0: the input impedance $Z_i = Z_0$, independent of the length l.

2. The line is short-circuited at $z = l$: the input impedance is then

$$Z_i = jZ_0 \tan \beta l$$ (10.70)

3. The line is open at $z = l$: the input impedance is then

$$Z_i = -jZ_0 \cot \beta l$$ (10.71)

4. The line is terminated at $z = l$ with an arbitrary impedance Z_l, but the line is very short compared with the wavelength, i.e. $\beta l = 2\pi l/\lambda \ll 1$, so that $\cos \beta l \approx 1$ and $\sin \beta l \approx 0$: the input impedance is then $Z_i \approx Z_l$.

A widely used type of impedance transformer works on the principles described above.

Let us suppose that we have a length of transmission line (labelled 2), terminated with its characteristic impedance Z_2, which is to be connected to a line 1 of characteristic impedance Z_1 so that line 1 is also free of reflections. How can this be done? Let us suppose that the solution could be found in the form of a line 3 of length l and characteristic impedance Z_3 placed between 1 and 2 (see Fig. 10.10).

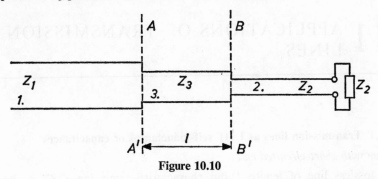

Figure 10.10

We already know that the input impedance at BB', looking to the right, is Z_2 (see example given under 1 above). What we want is that the impedance at AA', looking to the right, should be Z_1 so that waveguide 1 also has a matched termination. In general

$$Z^{AA'}_{\rightarrow} = Z_3 \frac{Z_2 + jZ_3 \tan \beta l}{Z_3 + jZ_2 \tan \beta l} \qquad (10.72)$$

If we now choose $\beta l = \pi/2$, so that $l = \lambda/4$ and $\tan \beta l \rightarrow \infty$, then it follows from equation (10.72) that

$$Z^{AA'}_{\rightarrow} = Z_3^2/Z_2$$

We wanted $Z^{AA'}_{\rightarrow} = Z_1$; the impedance transformer will therefore meet our requirements if

$$Z_3^2 = Z_1 Z_2 \quad \text{and} \quad l = \lambda/4 \qquad (10.73)$$

The above argument does not take parasitic capacitances into account. These can occur at the junction of two transmission lines of different dimensions (see Fig. 10.11 for an example in a coaxial cable). If the

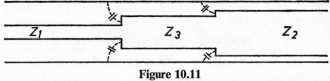

Figure 10.11

difference in diameters is not too great (i.e. Z_1 and Z_2 do not differ much), equation (10.73) remains a good approximation, but as the difference in diameter increases, the extra capacitances cause greater inaccuracy in equation (10.73). In practice, one still chooses $Z_3^2 = Z_1 Z_2$ in such cases, but the length is taken somewhat less than a quarter wavelength to compensate for the parasitic capacitances.

11 APPLICATIONS OF TRANSMISSION LINES

11.1 Transmission lines as UHF self-inductances or capacitances

Line with short-circuited end

A lossless line of length l and characteristic impedance Z_0 is short-circuited at $Z = l$ (see Fig. 11.1).

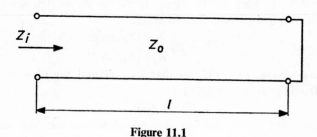

Figure 11.1

As we have shown in the previous section, the input impedance of this line is

$$Z_i = jZ_0 \tan \beta l$$

The right-hand side is imaginary, so Z_i is a reactance jX_i. If $\tan \beta l > 0$, Z_i is inductive; this occurs if

or

$$0 < \beta l < \pi/2, \qquad \pi < \beta l < 3\pi/2 \text{ etc.}$$
$$0 < l < \lambda/4, \qquad \lambda/2 < l < 3\lambda/4 \text{ etc.} \tag{11.1}$$

In this case, the input reactance of the line is equivalent to the impedance of a coil with self-inductance L, such that

$$j\omega L = jZ_0 \tan \beta l$$

or

$$L = \frac{Z_0}{\omega} \tan \beta l = \frac{Z_0}{\omega} \tan \frac{2\pi l}{\lambda} = \frac{Z_0}{\omega} \tan \frac{\omega l}{c} \tag{11.2}$$

Only for very small values of βl can tan βl be replaced by βl. In such cases, $L \approx Z_0 l/c$, i.e. independent of frequency, as in lumped self-inductances. In general, however, the equivalent, distributed self-inductance of a length of transmission line is frequency-dependent.

If equation (11.1) is not satisfied, i.e. if the line does not have one of the lengths given by the above formulae, tan $\beta l < 0$ and the input impedance is capacitive. The equivalent capacitance C is then given by

$$\frac{1}{j\omega C} = jZ_0 \tan \beta l$$

or

$$C = -\frac{1}{\omega Z_0 \tan \beta l} = -\frac{1}{\omega Z_0} \cot \frac{\omega l}{c} \tag{11.3}$$

In this case

$$\lambda/4 < l < \lambda/2, \ 3\lambda/4 < l < \lambda, \ \ldots \tag{11.4}$$

This equivalent C is again practically frequency-independent at values of l very near a quarter wavelength, but in general it is not frequency-independent, as may be seen from equation (11.3).

As the length of a short-circuited section of transmission line increases from zero, the input impedance at a given frequency is alternately inductive and capacitive (see Fig. 11.2).

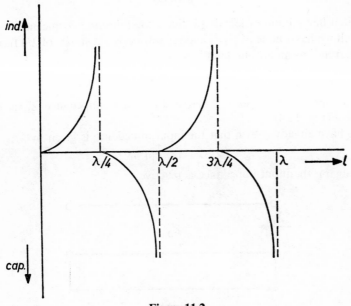

Figure 11.2

Apart from the frequency dependence, there is another striking difference between a normal lumped capacitance and this equivalent distributed capacitance. If the former is lossless, it has an infinite resistance for direct current. A short-circuited section of transmission line behaving as a capacitance, however, forms a short-circuit for direct current unless special measures are taken to prevent this.

Fig. 11.3 sketches the standing current and voltage waves for a short-

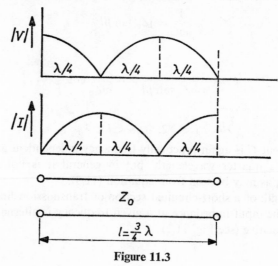

Figure 11.3

circuited line section of length $\frac{3}{4}\lambda$. Since the reflection coefficient $r = -1$, the voltage has a node at $z = l$ and an anti-node a distance of $\lambda/4$ further; the current has an anti-node at $z = l$.

Open-ended line

A lossless line of length l and characteristic impedance Z_0 is open at $z = l$ (Fig. 11.4).

We have already shown that the input impedance is given by

$$Z_i = -jZ_0 \cot \beta l$$

Here again, the input impedance is reactive.

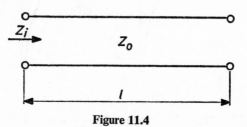

Figure 11.4

If equation (11.1) is satisfied, cot $\beta l > 0$, so that Z_i is capacitive. The value of the equivalent capacitance follows from

$$1/j\omega C = -jZ_0 \cot \beta l$$

or

$$C = \frac{1}{\omega Z_0} \tan \beta l \qquad (11.5)$$

In general, C is again frequency-dependent, apart from very low values of βl, where $C \approx l/Z_0 c$.

If cot $\beta l < 0$, so that Z_i is inductive (which occurs when l has the values given by equation 11.4), the equivalent self-inductance follows from

$$j\omega L = -jZ_0 \cot \beta l$$

or

$$L = \frac{-Z_0}{\omega} \cot \beta l \qquad (11.6)$$

In Fig. 11.5, the magnitude of the equivalent C or L is sketched for a given frequency as a function of the length of the transmission line.

A lumped coil has no resistance for direct current if the losses are zero. An inductance formed by a lossless open transmission line, however, has an infinite d.c. resistance.

The reflection coefficient of an open line is $r = 1$; the voltage wave has an anti-node at the open end and a node a quarter wavelength away, while the current wave has a node where the voltage wave has an anti-node and vice versa (this is represented in Fig. 11.6).

11.2 Transmission lines as resonant circuits

A resonant circuit for low frequencies in its simplest form consists of a capacitance and a self-inductance connected in series or in parallel. We saw in the previous section that a section of open or short-circuited transmission line can act both as a capacitance and as a self-inductance, depending on its length. It should thus be possible to make a resonant circuit at higher frequencies by replacing both the lumped C and L by a section of transmission line of suitable length and characteristic impedance. It goes without saying that this can be done in many different ways, and only a few alternatives will be discussed here.

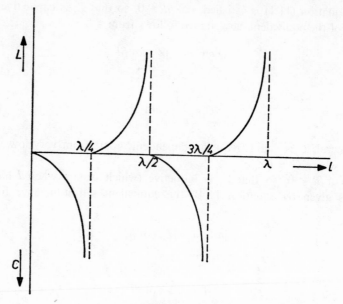

Figure 11.5

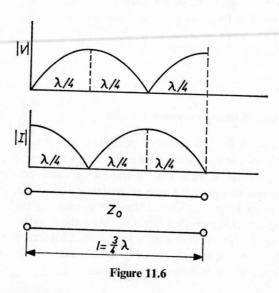

Figure 11.6

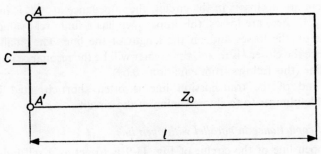

Figure 11.7

Lumped capacitance in parallel with short-circuited line

The circuit is shown in Fig. 11.7. The short-circuited line has to act as a self-inductance. Resonance will occur at ω_0, where

$$\frac{1}{j\omega_0 C} = -jZ_0 \tan \beta l$$

or

$$\frac{1}{\omega_0 C} = Z_0 \tan \beta l \qquad (11.7$$

This transcendental equation can in principle be solved (graphically) by values of l between 0 and $\lambda/4$, between $\lambda/2$ and $3\lambda/4$, etc. In general, however, we choose $l < \lambda/4$. In the first place, this makes the dimensions of the resonant system smaller, while it may also be of importance for other properties of the resonant circuit (e.g. bandwidth)—see Fig. 11.8.

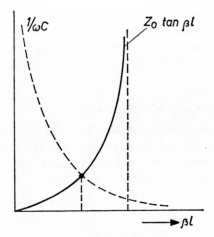

Figure 11.8

If there are no losses in the circuit, the impedance at AA' is infinite at $\omega = \omega_0$. If there are losses, this impedance has a finite real value, which depends on the losses and on the length of the line. The smaller C is, and hence the closer l is to $\lambda/4$, the greater will be the resonance impedance of the line (this follows from equation 10.69).

The end of the transmission line is often short-circuited with an adjustable plunger to facilitate tuning of the circuit.

Lumped capacitance in parallel with open line

If the open line of the circuit of Fig. 11.9 is to act as a self-inductance, its length should be slightly greater than a quarter wavelength. The resonance frequency ω_0 is given by

$$\frac{1}{\omega_0 C} = -Z_0 \cot \beta l \tag{11.8}$$

If circumstances make it desirable to have quite a large capacitance in the tuned circuit, and hence a low self-inductance, this solution may prove useful. The solution given in equation (11.7) may then be difficult to

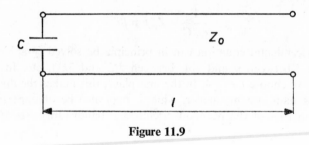

Figure 11.9

realize in practice if the transmission line has to be very short. An open line a little more than a half wavelength long may have the disadvantage that it can also give resonance at a frequency lower than the desired resonance frequency.

Short-circuited line in parallel with open line

Both the sections of transmission line shown in Fig. 11.10 have lengths l_1 and l_2 less than $\lambda/4$ and the same characteristic impedance Z_0. The short-circuited one is inductive, and the open one capacitive.

The resonance frequency follows from $Z^{AA'}_{\rightarrow} = -Z^{AA'}_{\leftarrow}$, i.e.

$$Z_0 \tan \beta l_1 = Z_0 \cot \beta l_2$$

or

$$\tan \beta l_1 . \tan \beta l_2 = 1 \tag{11.9}$$

Now

$$\tan(\beta l_1 + \beta l_2) = \tan \beta l = \frac{\tan \beta l_1 + \tan \beta l_2}{1 - \tan \beta l_1 \tan \beta l_2} \qquad (11.10)$$

where $l = l_1 + l_2$.

Substituting the resonance condition of equations (11.9) in (11.10), we obtain

$$\tan \beta l \rightarrow \infty$$

or

$$\beta l = \frac{\pi}{2}, \frac{3\pi}{2}, \text{ etc.} \qquad (11.11)$$

or

$$l = l_1 + l_2 = \frac{\lambda}{4}, \frac{3\lambda}{4}, \text{ etc.}$$

As was to be expected from the figure, the resonance frequency depends only on the total length. A short-circuited line of length $l = y/4$ behaves as a parallel tuned circuit with the resonance frequency

$$\omega_0 = 2\pi f_0 = 2\pi c / \lambda$$

In Fig. 11.11, the reactance of such a line is drawn in the neighbourhood of the resonance frequency.

Series circuit formed by a short-circuited line

In a series-resonant circuit with lumped elements, the impedance is capacitive for frequencies lower than the resonance frequency, zero for $\omega = \omega_0$ and inductive for $\omega > \omega_0$. If we want to reproduce this situation with a short-circuited transmission line, the line will have to be an even number of quarter wavelengths long, e.g. $l = \lambda/2$ (Fig. 11.12).

The resonance frequency is $\omega_0 = 2\pi c/\lambda = \pi c/l$.

The impedance of the 'circuit', i.e. the input impedance of the line, is $Z_i = jZ_0 \tan \beta l$ where $\beta l = \pi$, i.e. $Z_i = 0$ at resonance, $\beta l < \pi$, i.e. Z_i is capacitive for $\omega < \omega_0$ and $\beta l > \pi$, i.e. Z_i is inductive for $\omega > \omega_0$.

Parallel circuit formed by an open line

An open line of length $l = n\lambda/2$, where n is an integer, acts as a parallel tuned circuit, with as lowest resonance frequency

$$\omega_0 = \frac{2\pi c}{\lambda} = \frac{\pi c}{l}$$

Series circuit formed by an open line

An open line of length $l = (2n+1)\lambda/4$, where n is zero or an integer, acts as a series tuned circuit with resonance frequencies

$$\omega_0 = \frac{2\pi c}{\lambda} = \frac{2\pi c}{4l}(2n+1) = \frac{\pi c}{2l}(2n+1)$$

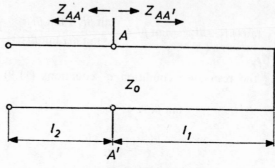

Figure 11.10

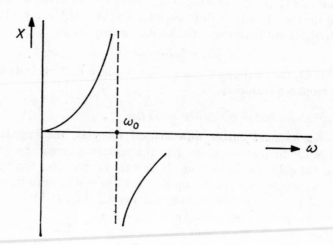

Figure 11.11

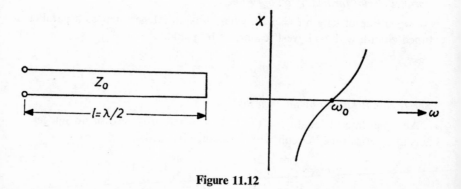

Figure 11.12

11.3 Transmission lines as wavemeters

The input impedance of a short-circuited section of transmission line, $jZ_0 \tan \beta l$, does not change when the length changes by an amount $n\lambda/2$, where n is an integer. This can be achieved by using an adjustable plunger for the short circuit. Now if a loop is attached to the open end as shown in Fig. 11.13, the line will resonate at a frequency ω_r given by:

$$\omega_r L = -Z_0 \tan \beta l = -Z_0 \tan \frac{\omega_r l}{c} \qquad (11.12)$$

where L is the self-inductance of the loop and l is the length of the line. There are a number of values of l which satisfy this equation; they occur at intervals of a half wavelength.

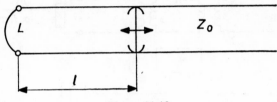

Figure 11.13

We can now measure the wavelength (or frequency), for example, of an oscillator by 'catching' part of the energy from the oscillator in the loop and measuring the distance between successive resonances with a centimetre scale on the transmission line. For this purpose, we need a detector in the loop. The rectified voltage is measured by an ammeter in series with a resistance at the 'dead' end of the line (resonance occurs when the deflection is maximum). In order to prevent short-circuiting of the detector, a capacitor should be included with such a high capacitance that it acts as a short-circuit for the frequencies to be measured (see Fig. 11.14).

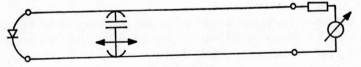

Figure 11.14

We shall now amplify on the above with a few general remarks about wavemeters and the measurement of wavelength or frequency.

Wavemeters are made in many different forms, all of which we shall not go into here. What all types have in common is a resonant circuit with as high a quality factor as possible, which can be tuned over the

range in question and which is coupled to the system whose frequency
is to be measured (for example, by means of a loop or holes in the
transmission line). In the microwave range, such a circuit is not composed
of lumped elements, but more generally uses a cavity resonator. (We shall
be discussing cavity resonators in more detail below.) Another possibility
is to use a length of transmission line as a wavemeter, as we have just seen.

Cavity resonators can in general be used in two different ways: as a
transmission-type meter and as a reactance-type. Fig. 11.15 shows the

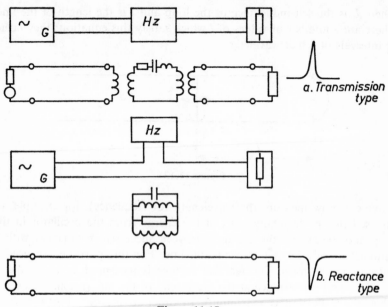

Figure 11.15

principle of the circuits in block-diagram form with the equivalent
electrical circuits (in which the cavity resonators are shown as low-
frequency resonant circuits).

If the cavity resonators are tuned to the frequency to be measured,
the output voltage across the load will pass through a peak (or a minimum)
as the generator frequency passes the frequency to be measured. As the
quality factor of the resonator is higher, the peak (or minimum) will be
sharper, and it will be possible to determine the frequency more accurately.
The cavity resonators should also be reasonably free of resonances in
frequency ranges outside the measuring range, since to ensure that the
results are unambiguous we must have no undesirable extra resonances
in the frequency range under investigation, such as those from higher
modes in the field pattern in the cavity resonator. The accuracy is also

limited by temperature and humidity variations, and of course by the mechanical accuracy of the scale on the tuning system.

These cavity resonators must be suitably calibrated before use, since they are not primary wavelength standards, but secondary ones. This is often done by the use of quartz-crystal oscillators, whose resonant frequency is reduced so far with the aid of frequency dividers that it can be determined by accurate time measurements and a counting mechanism.

It is possible to use a transmission-type wavemeter for frequency measurements without having all the energy flowing through the wavemeter, through the use of a directional coupler (a component which will be described below, and which allows part of the energy flowing through it to be tapped off into a side line). This system is illustrated in Fig. 11.16.

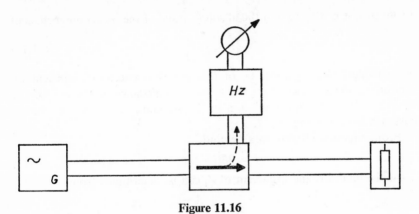

Figure 11.16

11.4 Impedance measurements with transmission lines

We have already shown that the voltage distribution along a finite transmission line depends on the terminal impedance Z_l (see equation 10.43). If we measure the voltage distribution along the line, we can deduce the terminal impedance from it. It is also possible to determine the reflection coefficient from the voltage distribution, and then to calculate the terminal impedance from the reflection coefficient by means of equations (10.42) and (10.50).

The voltage along the line can be measured with a 'standing-wave detector', an adjustable probe coupled to a detector. The general form of this voltage is illustrated in Fig. 11.17.

Firstly we measure the voltage standing-wave ratio (equation 10.64)

$$\eta = \frac{|V|_{max}}{|V|_{min}} = \frac{1+|r|}{1-|r|}$$

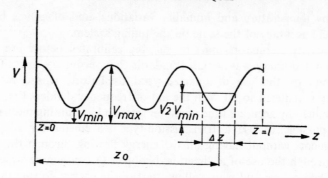

Figure 11.17

From this, it is easy to find the absolute value of the reflection coefficient

$$|r| = \frac{\eta - 1}{\eta + 1} \tag{11.13}$$

We calculate the argument φ_r of the complex reflection coefficient by measuring the distance from the terminal impedance at the end of the line $(z = l)$ to the first minimum in the standing wave pattern, i.e. the distance $(l - z_0)$ in Fig. 11.17.

From equation (10.60), we now find

$$\varphi_r = 2\beta(l - z_0) - \pi \tag{11.14}$$

Now the modulus and argument of Z_l can be determined from

$$Z_l = Z_0 \frac{1 + r}{1 - r}$$

by using equations (11.13) and (11.14).

An alternative, accurate method for the determination of $|r|$ (and also of the position of the minimum) is to measure the minimum Δz at the 3 dB points of the voltage standing wave. It follows from the data in Section 10.3 that

$$\sqrt{2} = \frac{\sqrt{1 + |r|^2 - 2|r| \cos \beta \Delta z}}{1 - |r|} \tag{11.15}$$

The value of $|r|$ can thus be calculated from the measured value of Δz. If $\Delta z \ll \lambda$, $\cos \beta \Delta z$ may be written

$$\cos \beta \Delta z = 1 - (\beta \Delta z)^2 / 2$$

so that to a first approximation

$$\beta \Delta z \approx \frac{1 - |r|}{\sqrt{|r|}} \tag{11.16}$$

It is rather time-consuming to have to calculate the terminal impedance Z_l

from $Z_l = Z_0(1+r)/(1-r)$ each time $|r|$ and φ_r are determined by measurement, so that in practice the Smith diagram (see Appendix 2) is often used, in which the reflection is plotted in polar coordinates in the complex plane and in which the lines are drawn for constant values of R_l/Z_0 and X_l/Z_0, where

$$Z_l = R_l + jX_l \tag{11.17}$$

$$\frac{R_l}{Z_0} + j\frac{X_l}{Z_0} = \frac{1+r}{1-r} = \frac{1+|r|\,e^{j\varphi_r}}{1-|r|\,e^{j\varphi_r}} \tag{11.18}$$

The loci of r for constant R_l/Z_0 and X_l/Z_0 are found to be families of circles in the complex r plane.

Since $|r| \leqslant 1$, all the values of R_l/Z_0 and X_l/Z_0 which exist lie within the unit circle in this complex plane. Taking all the impedance values relative to Z_0 makes the diagram suitable for use in transmission lines with any value of Z_0.

The diagram is not suitable for extreme values of R_l/Z_0 and X_l/Z_0; the accuracy is low for values lower than $0{\cdot}1$ and higher than 10. In these regions, small changes in $|r|$ and φ_r, corresponding to short distances in the diagram, give rise to such large changes in R_l/Z_0 and X_l/Z_0 that small errors in measurement can lead to large errors in R_l and X_l.

11.5 Transmission lines as impedance transformers

We have seen that the input impedance of a transmission line of characteristic impedance Z_0, terminal impedance Z_l and length l is given by

$$Z_i = Z_0\frac{Z_l + jZ_0\tan\beta l}{Z_0 + jZ_l\tan\beta l} \tag{11.19}$$

In other words, the transmission line of length l transforms the impedance Z_l at $z = l$ to Z_i, at $z = 0$, in accordance with the above formula.

We have already encountered a simple application of this principle in the 'quarter-wave transformer' described in Section 10.4: if $\beta l = \pi/2$, $3\pi/2$, etc., or l is an odd multiple of $\lambda/4$, Z_l is transformed to $Z_i = Z_0^2/Z_l$. This arrangement is often used to 'match' Z_l with Z_i, i.e. to provide the line with a reflection-free termination.

An even simpler example occurs where l is a whole number of half wavelengths, i.e. βl is a whole number times π. Now $\tan\beta l = 0$ and $Z_i = Z_l$, and if the line is lossless this is an ideal transformer with a transformation ratio of one, independent of the characteristic impedance.

It can be shown in general that if the length of the line varies with a given value of Z_l, the input impedance describes a circle in the complex plane, symmetrical with respect to the real axis (see Fig. 11.18).

Every time l increases by $\lambda/2$, i.e. βl increases by π, the end-point of the vector Z_i travels once round the circle. The points of intersection with the real axis are denoted by R_1 and R_2. With special values of l, the line

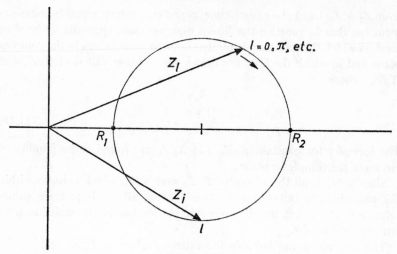

Figure 11.18

transforms the complex impedance Z_l to a real input impedance R_1 or R_2.

In order to show that Z_i really does describe a circle as shown in Fig. 11.18, we must write equation (11.19) in another form. Making use of Euler's equation and the expression for the reflection coefficient, we derive from equation (11.19)

$$\frac{Z_i}{Z_0} = \frac{Z_l(e^{j\beta l} + e^{-j\beta l}) + Z_0(e^{j\beta l} - e^{-j\beta l})}{Z_0(e^{j\beta l} + e^{-j\beta l}) + Z_l(e^{j\beta l} - e^{-j\beta l})}$$

$$= \frac{1 + \dfrac{Z_l - Z_0}{Z_l + Z_0} \cdot e^{-2j\beta l}}{1 - \dfrac{Z_l - Z_0}{Z_l + Z_0} \cdot e^{-2j\beta l}} = \frac{1 + |r|\, e^{-2j\beta l + j\varphi_r}}{1 - |r|\, e^{-2j\beta l + j\varphi_r}}$$

$$= \frac{1 + |r|^2}{1 - |r|^2} + \frac{2|r|}{1 - |r|^2}\, e^{j\varphi} \qquad (11.20)$$

where

$$\varphi = \varphi_r - 2\beta l + \tan^{-1} \frac{|r|\sin(\varphi_r - 2\beta l)}{1 - |r|\cos(\varphi_r - 2\beta l)} \qquad (11.21)$$

This means that the input impedance can be written in the form

$$\frac{Z_i}{Z_0} = a + b\, e^{j\psi} \qquad (11.22)$$

where a, b and ψ are real, and a and b are independent of l. Equation (11.22) represents a circle of radius b, with its centre on the real axis a distance a from the origin.

The points R_1 and R_2 are now easily found from equation (11.20)

$$\frac{R_1}{Z_0} = \frac{1-|r|}{1+|r|} \quad \text{and} \quad \frac{R_2}{Z_0} = \frac{1+|r|}{1-|r|} \tag{11.23}$$

since φ has the values π and 0 respectively at these points. It follows further from equation (11.23) that $R_1 R_2 = Z_0^2$.

We shall now describe a few more impedance transformers made of transmission lines, without dealing with the subject exhaustively. Firstly, see the transformer in Fig. 11.19. The resistance R_l has to be transformed to the value Z_0 at AA'. It would be possible to do this with the quarter-wave transformer discussed above, but this would mean using a piece

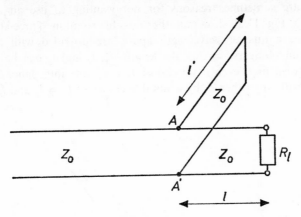

Figure 11.19

of line of a different characteristic impedance. It is also possible to obtain this transformation with a line of the same impedance. An extra piece of short-circuited line of length l' is attached to the line at a distance l from R_l. This is fixed at right angles to the main line, and is called a 'stub'. The value of l is chosen so that the real part of the input admittance (admittance is the reciprocal of impedance) of the line section l terminated with R_l is equal to $1/Z_0$, as seen from AA'. The imaginary part of the input admittance can be neutralized by suitable choice of l', since the stub gives a reactive input admittance. After some manipulation of the formula for the input impedance of a finite line, we find the following expressions for l and l'

$$\tan \beta l = \sqrt{\frac{R_l}{Z_0}} \quad \text{and} \quad \tan \beta l' = \sqrt{R_l Z_0}/(R_l - Z_0) \tag{11.24}$$

It may be seen from these formulae that any value of R_l can be matched with any value of Z_0 by a suitable choice of variables l and l'. As βl or $\beta l'$ passes from 0 to π, the value of the tangent passes from $-\infty$ to $+\infty$.

It must thus be possible to adjust these lengths by at least a half wavelength. For l' this can be done with an adjustable short-circuit plunger; for l, we must design a line section which can be varied over at least a half wavelength without changing the characteristic impedance. (Such a construction is variously named 'trombone', 'phase shifter', 'adjustable line length' or 'line stretcher'.)

Matching can be carried out in a similar way when the terminal impedance is not real but complex.

Many other types of impedance transformers are possible. For example, one described in Philips Research Reports (**3**, 140, 1948) is made of a length of screened parallel-wire transmission line.

There are sometimes reasons for not wanting to use an adjustable line length. Fig. 11.20 shows another possible solution. Three stubs, fixed to the line a quarter wavelength apart, are provided with adjustable short-circuit plungers, so that the lengths l_1, l_2 and l_3 can be varied. It can be shown that, even if we choose $l_1 = l_3$, any impedance Z_l can be matched with any other Z_i by a suitable choice of $l_1 = l_3$ and l_2.

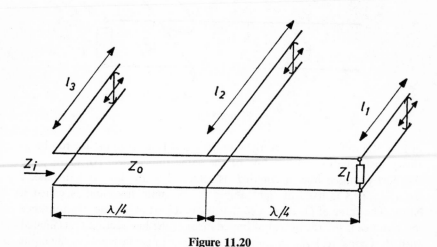

Figure 11.20

It should be mentioned in passing that this impedance transformer and those mentioned above, as well as those which we shall discuss below, can only be used over a narrow frequency range. All kinds of frequency-dependent distances play a role in the transformation process, so that the quality of the matching steadily deteriorates as we get further from the frequency for which the transformer was designed; in any case, the impedance range which can be covered by the transformation will be reduced.

Fig. 11.21 illustrates a slightly different type of impedance transformer, the 'slug tuner', whose operation is based on the presence of movable pieces of dielectric (the slug tuner shown in Fig. 11.21 is a coaxial version). Two movable rings of dielectric material (dielectric constant ε, with low losses), each a quarter wavelength long (with the wavelength measured in the dielectric, i.e. $\lambda = \lambda_0/\sqrt{\varepsilon}$), are placed in a coaxial cable. The distance S in the figure can be varied, while the two pieces of dielectric remain symmetrical with respect to AA'; or the whole system can be moved along the cable, with S constant.

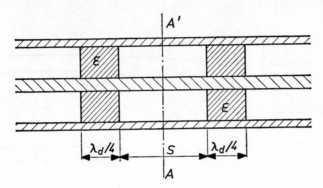

Figure 11.21

When $S = 0$, the line contains a piece of dielectric half a wavelength long; in other words, the line contains an intermediate segment of characteristic impedance Z_0' and length $\lambda/2$. The impedance transformation ratio of such a segment is 1, independent of the value of $Z_0' = Z_0/\sqrt{\varepsilon_r}$.

If we now allow S to increase from zero, we get two quarter-wave transformers connected by a line segment of length S. The maximum total transformation occurs when $S = \lambda/4$; the amplitude of the transformation ratio is then ε_r^2. This means that impedances with a value up to $\varepsilon_r^2 Z_0$ can be transformed to Z_0 with this system. Regarding the phase of the transformation, when S remains constant the amplitude of the transformation ratio does not change; as the system is moved along the coaxial cable, therefore, only the phase varies. The modulus and argument of the reflection coefficient can thus be varied more or less independently of one another (changing S without moving the centre of the system has little effect on the phase). Parasitic reflections have not been taken into account here.

We shall close this section with a few remarks on tapered sections (gradual transitions between two different segments of transmission line).

One of the simplest methods of connecting two lines of different characteristic impedance so as to avoid reflections in a given frequency range is to make the transition between the two lines gradual (see Fig. 11.22). If the transition region (of length d) between the characteristic impedances Z_1 and Z_2 is only made long enough with respect to the wavelength, the change in impedance will be so gradual that few reflections will occur. It would, of course, be better merely to use the shortest possible transition length for a certain maximum permissible value of the reflection. This is not a simple problem in principle, since we should really also determine the optimum form of the transition (linear as shown in Fig. 11.22, exponential, or some other form).

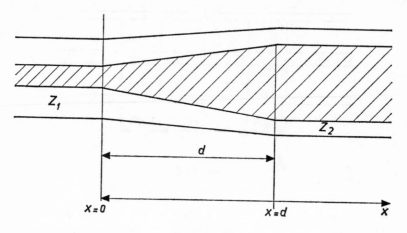

Figure 11.22

The problem may be clarified by the following example. We recall that the equations for the transmission line are

$$\frac{dV}{dx} = -ZI \quad \text{and} \quad \frac{dI}{dx} = -YV$$

where Z and Y are respectively the impedance $(R+j\omega L)$ and the admittance $(G+j\omega C)$ of the line, both per unit length. In this situation, Z and Y are no longer independent of the position x. For example, by eliminating I we find

$$\frac{d^2V}{dx^2} - \frac{1}{Z(x)} \frac{dZ(x)}{dx} \frac{dV}{dx} - Z(x)Y(x). \quad = 0$$

The problem now is to find the optimum forms of $Z(x)$ and $Y(x)$ for this situation. It is not easy to provide an exact solution to this problem;

various approximate solutions have been given in textbooks and in the periodical literature. If the transition is linear, it must be at least one wavelength long at the frequency in question, and preferably several wavelengths. The greater the difference between Z_1 and Z_2, the longer must be the transition if the reflections are to be kept down to a specific maximum permissible level.

If the form of the transition is such that $\ln Z_0$ is proportional to x in the transition region, where $Z_0(x)$ is the characteristic impedance at x, then the modulus of the reflection coefficient with air as the dielectric is

$$|r| \approx \tfrac{1}{2} \ln \frac{Z_1}{Z_2} \frac{|\sin \theta|}{\theta}$$

where $\theta = 2\pi d/\lambda$, the electric length of the transition.

11.6 Discontinuities

We have already mentioned that parasitic capacitances can occur (for example, in impedance transformers), but so far we have ignored them. We shall now devote some attention to this subject without going into full details, since a complete treatment of the subject is beyond the scope of this book.

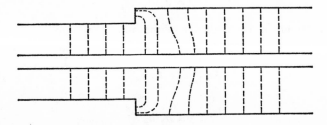

Figure 11.23

Fig. 11.23 shows an example of a coaxial cable with a discontinuity in its diameter. The approximate form of the electric lines of force in the neighbourhood of the discontinuity is indicated. All the lines of force end at right angles to the highly conductive metal parts. Due to the form of the discontinuity, and hence of the lines of force, the electric field must have an axial component in this region. Since the simplest mode (the TEM mode)—the only one we have been considering so far—only

contains transverse components, higher modes must occur in the neighbourhood of such a discontinuity, and part of the energy of the principal
mode will be transformed into higher modes. If the dimensions of the
transmission line are such that higher modes can be propagated along the
line, the discontinuity will give rise to a permanent 'mode conversion'.
If only the TEM mode can be propagated (see equation 8.17), the higher
mode will disappear within a very short distance. All the energy will then
be converted back to the principal mode (if we neglect line losses) and the
field pattern at some distance from the disturbance reflects the normal
TEM picture again.

The effect of such a discontinuity, at such a distance that the field
pattern has returned to normal and in the absence of any further discontinuities nearby, can be taken into account in the equivalent circuit of the
line by the inclusion of an extra admittance, in the present case a capacitance, $Y_d = j\omega C$ (see Fig. 11.24). The value of C follows from the magnitude of the sudden change in diameter.

Figs. 11.25–11.27 show further cases of abrupt changes in diameter
of the conductors of a coaxial line, with indications of the approximate
value of the parasitic capacitance in the various cases.

In Fig. 11.25, the parasitic capacitance shown, C_1, which is due to the
discontinuity indicated in the diameter of the inner conductor, is given by

$$C_1 = 2\pi r_1 \varepsilon_{r_2} C_d(\alpha) \tag{11.25}$$

where

$$\alpha = a_1/b_1 \tag{11.26}$$

The discontinuity in the diameter of the outer conductor of the coaxial
line shown in Fig. 11.26 gives rise to a parasitic capacitance C_2, such that

$$C_2 = 2\pi r_2 \varepsilon_{r_2} C_d(\alpha) \tag{11.27}$$

where

$$\alpha = a_2/b_2 \tag{11.28}$$

Finally, Fig. 11.27 shows a case where the diameters of both the inner
and the outer conductor change suddenly. The parasitic capacitance
indicated in this case, C_3, is given by

$$C_3 = 2\pi r_1 \varepsilon_{r_1} C_\alpha(\alpha) \tag{11.29}$$

where

$$\alpha = a_1/b_1 \tag{11.30}$$

The total extra capacitance in this case is $C_c = C_2 + C_3$.

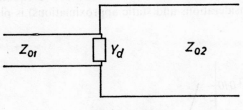

Figure 11.24

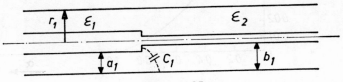

Figure 11.25

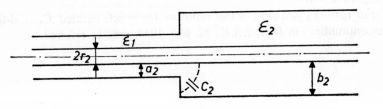

Figure 11.26

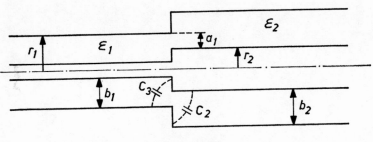

Figure 11.27

The quantity $C_d(\alpha)$ which occurs in the above equations (based on theoretical considerations and static approximations) is plotted in pF/cm in Fig. 11.28.

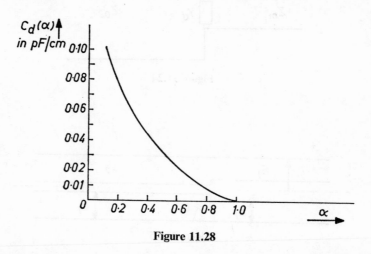

Figure 11.28

If the diameter of each coaxial cable changes by a factor of not more than five, the result given by these formulae is accurate within about 20% for $\alpha > 0\cdot1$.

For further discussion of this problem, the article entitled 'Coaxial-line discontinuities', in *Proc. I.R.E.*, **32**, 695, 1944, may be referred to.

12 COMPONENTS IN PARALLEL-WIRE TRANSMISSION LINES

In this section we shall review a variety of the components widely used in parallel-wire lines. The examples will be restricted to coaxial and Lecher lines.

12.1 Supports

In coaxial lines with air as dielectric, the inner conductor must often be supported at one or more points between the two connecting plugs. The support should be made of an insulating material with very low losses. The presence of such a disc of material (see Fig. 12.1) will increase the

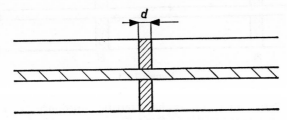

Figure 12.1

characteristic impedance at that point because the capacitance per unit length increases. It is therefore desirable to keep the dielectric constant down, which also reduces any reflections that might occur. The effect of the support can be represented by an extra capacitance C at the site of the support. If the thickness $d \ll \lambda$, we may assume that the capacitance is concentrated at this point. Its effect can be cancelled out over a limited frequency range by the inclusion of a short-circuited length of line as shown in Fig. 12.2.

If there are standing waves in the original coaxial line, the effect of the support can be reduced still further by, where possible, placing it at a node of the voltage wave.

119

If it is desirable for mechanical reasons to have more than one support, these can be placed at such a distance from one another that they neutralize one another's effect as far as possible (see Fig. 12.3). If C is the capacitance of one support, the distance a should be such that

$$Z_0 \tan \beta a = 2/\omega C \tag{12.1}$$

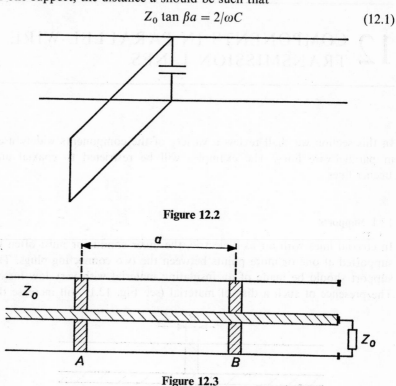

Figure 12.2

Figure 12.3

If the line is terminated with its characteristic impedance Z_0, it will have a travelling wave outside the region between the two supports; between the supports, there will be a standing wave (see Fig. 12.4).

It is also possible to keep the characteristic impedance constant at the point where the supports are placed by making the internal diameter of the outer conductor slightly larger here. However, this is not effective in reducing reflections, especially if the extra capacitance introduced by the supports tends to be large. A similar result can be obtained by reducing the diameter of the inner conductor somewhat.

Supporting the inner conductor by means of a $\lambda/4$ side-arm (Fig. 12.5) is another possible solution which gives low reflections in a narrow frequency range. Alternatively, the support can be made a half-wavelength thick, so that the transformation ratio in a limited range of frequencies is independent of Z_0.

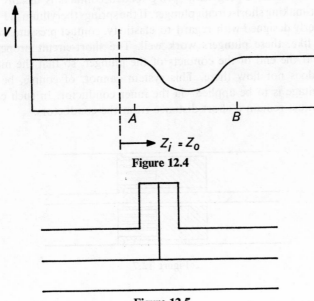

Figure 12.4

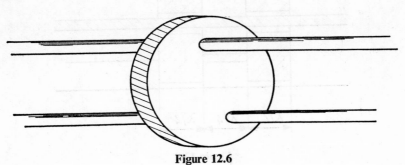

Figure 12.5

12.2 Short-circuit plungers

We have already mentioned adjustable short-circuits in parallel-wire lines, but have not gone into detail about their nature.

In Lecher lines, simply connecting the two conductors will not form an adequate short-circuit at high frequencies, as the electromagnetic field will simply 'flow round' (cf. field distribution of Fig. 8.2). It is better to use a disc short-circuit, as shown in Fig. 12.6; the disc should have a sufficiently large diameter. It sometimes happens, for example, that a d.c. voltage has to be applied between the two conductors. The disc should then be made in two parts, separated by a very thin insulating layer.

Figure 12.6

In coaxial lines, a plug with spring-loaded contacts is usually used as a contact-making short-circuit plunger. If the springs (beryllium in Fig. 12.7) are properly designed with regard to elasticity, contact pressure, soldering and the like, these plungers work well. The short-circuit proper is not situated at the end of the contacts of the plunger, so that the maximum current does not flow there. This system cannot, of course, be used if a d.c. voltage is to be applied via the inner conductor. In such cases the plunger must remain a short distance from the conductors.

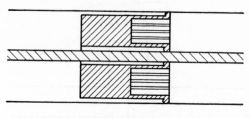

Figure 12.7

Unless special measures are taken, the high-frequency field leaks past the plunger along the conductors, so that the short-circuit is insufficient. Choke plungers (Fig. 12.8) are often used to avoid this problem. The plunger body here consists of three parts, each a quarter wavelength long and each forming a coaxial line, the characteristic impedance of which is alternately fairly high and low. Various kinds of folded designs are also possible, one of which is illustrated in Fig. 12.9.

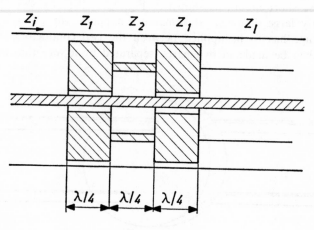

Figure 12.8

In order to understand the operation of choke plungers, we apply the formula for the impedance transformation of a quarter-wave transformer, as given above, a number of times. As we have seen, a line segment of length $\lambda/4$ and characteristic impedance Z_0 transforms the load impedance Z_l to $Z_i = Z_0^2/Z_l$. If $Z_l \neq 0$, but is small, then $Z_i \neq \infty$, but will be large.

In the case of Fig. 12.8, then

$$Z_i = Z_1^4/(Z_2^2 . Z_l) \qquad (12.2)$$

For not too extreme values of the impedances, for example, $Z_1 = 5$, $Z_2 = 40$ and $Z_l = 200$ ohms, which gives $Z_i = 1/500$ ohm. In general, the ratio Z_1/Z_2 should be small if the short-circuit is to be effective.

The plungers are often centred with teflon plugs.

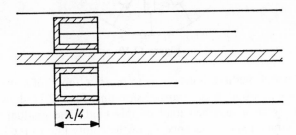

Figure 12.9

12.3 Matched loads

We have already seen that a transmission line terminated by its characteristic impedance does not give standing waves. In this case, the line is said to have a matched load. Matched loads are often required in microwave techniques, especially for making certain measurements. There is no need for the matched termination to transmit the power incident on it, as long as the power is absorbed without reflection (reflection coefficient 0, voltage standing-wave ratio 1). It is often possible to specify how closely the termination should approach this ideal behaviour for a given application, by stating that the voltage standing-wave ratio should be 1·01, 1·05, 1·10 or some similar value. Now it will be clear from what we have said above about transmission lines and the fields in them that there is no point in trying to provide, say, a coaxial line with a matched termination by soldering a normal resistance of Z_0 ohm, as used in low-frequency techniques, between the inner and the outer conductor. The wave impedance of such an arrangement will not be Z_0 ohm, since much of the field will flow round the resistance. A closer approximation will be obtained if a large number (n) of resistances of nZ_0 ohm each are connected in parallel between the inner and outer conductors in a given cross-section of the coaxial line (Fig. 12.10). As long as the demands on matching are not too stringent, this method can be used in practice.

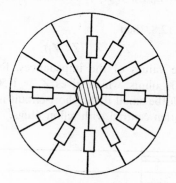

Figure 12.10

A much better method, however, is to insert a 'taper' of some high-loss material such as wood in the transmission line. Fig. 12.11 shows a transmission line with a matched load of this kind. The gradual transition from the characteristic impedance Z_0 of the empty line to the impedance Z_0' of the line filled with the high-loss material gives matching with very little reflection. The high losses in the material (large α) attenuate the waves; the line can be closed off with a metal plate behind the taper to prevent leakage of the residual field. If the voltage standing-wave ratio of the taper is to be less than $1 \cdot 01$, it must provide an attenuation of 23 dB (200-fold reduction in energy). It is further assumed that no other reflections come from any other source (for example, from the taper itself).

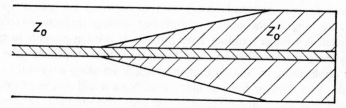

Figure 12.11

If high demands are made on the quality of the matching, the length of the taper will often be made greater than that given in Section 11.5 for the impedance transformation in question.

The type of termination used will, of course, depend on the power to be handled. The example given above is not very suitable for high powers, when it is preferable to use the construction shown in Fig. 12.12 since this makes it easier to cool the termination (e.g. with cooling fins or a liquid coolant). The latter termination works on a similar principle to the former. The termination is generally tested experimentally before use, and if necessary adjusted by an iterative or trial and error procedure.

A high-loss cable (with the same characteristic impedance as the line to be terminated) often provides quite a good matched load.

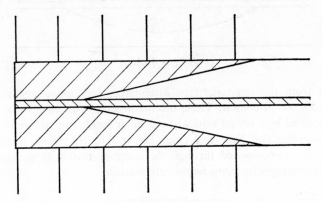

Figure 12.12

12.4 Attenuators

An attenuator in a transmission line works on much the same principle as the matched load discussed in the previous section. If we restrict ourselves for the moment to attenuators with a fixed attenuation (variable attenuators merely require an extra mechanism to adjust the coupling between the attenuator discussed below and the line), one possible solution is to use a section of transmission line with the inner conductor made of a high-loss material, or of a good conductor with a coating of a high-loss material. The damping produced by such an arrangement can be measured, and adjusted to the required value by trial and error. If the reflection coefficient is to be kept low, we cannot obtain losses of more than about 8 to 10 dB in this way. Higher values of attenuation can be obtained by inserting a double taper of high-loss material into the line in order to keep the reflections low (Fig. 12.13). The attenuation can be varied by varying the length of the taper, the exact lengths required being found by trial and error.

5*

An attenuator of this kind can be made adjustable by means of a device which allows the high-loss material to be introduced a variable distance into the line through a slit in the outer conductor. Since the transition, though gradual, is of finite length, the reflection coefficient of such an attenuator can be kept below certain limits only in a restricted frequency range.

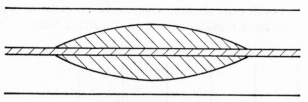

Figure 12.13

Apart from the resistance-type attenuators mentioned above, cut-off attenuators are also used. These often consist of two lengths of coaxial line connected by a round tube without inner conductor, where the length of the coupling can be varied (Fig. 12.14). The transverse wave in the coaxial line is propagated through the coupling in which the inner conductor is missing with considerable attenuation.

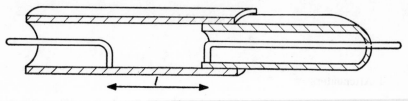

Figure 12.14

Apart from the low-attenuation range, when the coupling loops are very near one another, the attenuation (in dB) varies almost linearly with the length of the coupling (see Fig. 12.15). It is difficult to adjust this attenuator accurately in the low-attenuation region.

12.5 Detectors

We can only deal with this subject very briefly in this work. More detailed material is available in other literature.

Apart from various modifications necessitated by the shape and size of the components involved, the detection of microwaves is carried out in much the same way as in low-frequency electronics. However, certain

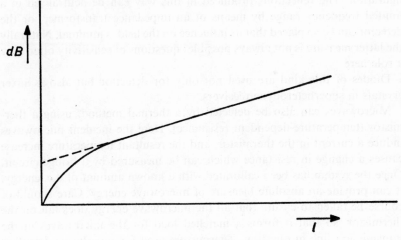

Figure 12.15

non-linear components such as vacuum diodes are not used, because of transit time effects at very high frequencies. Microwaves can be detected electrically by allowing the waves to fall on a crystal diode (Ge, Si, etc.) and measuring the induced current or voltage. Both point-contact diodes and junction diodes are used, in various forms (Fig. 12.16 shows the cartridge and coaxial types). The diodes are mounted in a holder in the

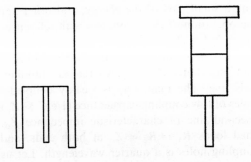

Figure 12.16

transmission line, in such a way that part of the field in the line falls on the non-linear element and is detected. Measures must be taken to ensure that the d.c. current produced can flow, while the diode housing does not short-circuit the high-frequency waves either galvanically or through parasitic capacitances. An element of this kind will often alter the field configuration. The reflections produced in this way can be neutralized in a limited frequency range by means of an impedance transformer, or the detector can be so placed that its influence on the field is minimal. Naturally the latter measure is not always possible; questions of sensitivity often play a role here.

Diodes of this kind are used not only for detection but also as mixer crystals in superheterodyne receivers.

Microwaves can also be detected by a thermal method, using a thermistor (temperature-dependent resistance). Here the incident microwaves induce a current in the thermistor, and the resultant temperature increase causes a change in resistance which can be measured by a bridge circuit. Once the system has been calibrated with a known amount of d.c. energy, it can provide an absolute measure of microwave energy. Care should of course be taken to ensure that all the microwave energy does fall on the thermistor, so that it forms a matched load for the microwaves in the transmission line in question. Thermistors used for microwave detection are made in the same form as the electrical detectors mentioned above.

12.6 Directional couplers

Directional couplers are widely used for the rapid but not too accurate measurement of reflection coefficients or voltage standing-wave ratios, the matching of transmission lines with the aid of suitable impedance transformers, checks on the operation of a microwave source by tapping off and detecting a small part of the energy, and other applications. We shall discuss the operation of these couplers with reference to the coaxial line version shown in Fig. 12.17.

The main transmission line 1 of characteristic impedance Z_0 is terminated with an impedance $Z_l \neq Z_0$, so that a standing wave will be produced in this line. The main line is weakly coupled—for example, by two small holes of low coupling capacitance ($1/\omega C \gg Z_0$) in a common wall—with a second line of characteristic impedance Z_0'. This second line has matched loads $R_1 = R_2 = Z_0'$ at both ends, and the distance between the coupling holes is a quarter wavelength. Let us imagine that the standing wave in the main line is separated into a wave travelling to the right and one travelling to the left and first consider the voltage wave travelling to the right. The voltage at A is V_A and that at B is V_B. V_B has a phase lag of 90° with respect to V_A, while their amplitudes will be the same if we disregard losses. A small part of the energy will be fed into the side line at both A and B (if the coupling is properly designed, the amount of energy will be the same in both cases). These waves will be propagated symmetrically in both directions in the side line until

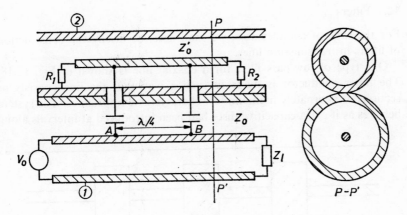

Figure 12.17

they are absorbed in R_1 or R_2 (or detected, by means not shown in the figure). The part of the main wave coupled into the side line at B and travelling to the left will arrive at A with a phase lag of $180°$ ($\lambda/2$) with respect to the part of the wave coupled into the side line directly at A. The two waves will thus interfere and cancel out. The wave coupled into the side line at A and travelling to the right is in phase with that coupled in at B when the former reaches B. In other words, the wave travelling to the right in the main line does not deliver any energy to R_1, but only to R_2.

The situation with a wave travelling to the left in the main line is just the reverse. Measuring the reflection coefficient or the voltage standing-wave ratio in the main line thus consists of measuring the voltages across R_1 and R_2 and determining their ratio, for example, with a 'ratiometer'. This method provides continuous information about the reflection conditions (the deflection of the meter across R_1 is minimum when reflection is minimum), unlike the method using the standing-wave detector discussed in Section 11.4. Sometimes only one of the matched loads in the side line is designed to have access to it and this can, for example, be used for monitoring a source.

Systems with various coupling factors, e.g. 3, 10, 15, 20 dB and other values, are available commercially. The quality reduces as the operating frequency deviates more from the nominal frequency: the electrical distance between the coupling holes is frequency-dependent, and causes the directional sensitivity to fall off as the frequency departs from the nominal value.

12.7 **Filters**

For the sake of completeness, we shall describe briefly the construction of filters in transmission lines.

One type of low-pass filter for a coaxial line is shown in Fig. 12.18. The inner conductor is made in such a way that the line consists of sections of alternately high and low characteristic impedance. The system behaves as if extra capacitive discs had been introduced at intervals along

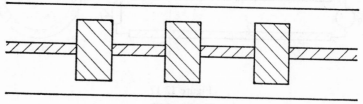

Figure 12.18

the line. At low frequencies, these can be regarded as lumped capacitances, but at high frequencies the capacitance is determined by their role as short line segments of low-characteristic impedance. The equivalent circuit is illustrated in Fig. 12.19, where it will be seen that this is a typical low-pass filter. The self-inductances are formed by the line segments of high impedance.

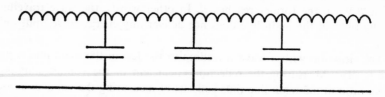

Figure 12.19

A high-pass filter can be introduced into a coaxial line as in Fig. 12.20. Short line segments ($< \lambda/8$) parallel to the main line represent the self-inductances of the equivalent circuit drawn in Fig. 12.21. Insulating dielectric plugs (ε) are placed in the inner conductor as series capacitances. The equivalent circuit is that of a high-pass filter. A modification of the system of Fig. 12.20, in which the parallel line segments are a quarter wavelength long, gives a band-pass filter. The capacitive plugs in the inner conductor can then be omitted.

The design of such filters goes beyond the scope of this book. For details, the reader is referred to J. J. Karakash, *Transmission Lines and Filter Networks* (Macmillan).

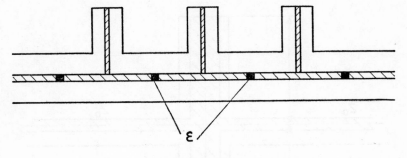

Figure 12.20

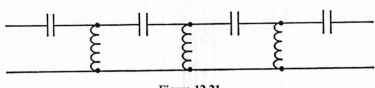

Figure 12.21

12.8 Power dividers

We shall discuss power dividers in greater detail, not because they are so widely used, but because they will increase our insight into the general use of discrete line segments (of length $\lambda/4$ and $\lambda/2$) in transmission-line systems.

A power divider is used for separating a given amount of power in a transmission line into two or more parts, which may or may not be equal.

We shall start by discussing one version (Fig. 12.22) for dividing the power into two equal parts. This circuit works well over a wide frequency range. We assume that the source in the main line (shown vertically) and the loads in the side-arms are matched to the lines in which they are situated. At the junction, the two side-arms are in parallel with a section of short-circuited line of length $l = \lambda/4$. The characteristic impedance of the short-circuited line segment is equal to that of the quarter-wave transformer from the input line to the junction. It will be clear from considerations of symmetry, quite apart from the differences in diameter, that equal amounts of energy will flow in the two side-arms. However, a large proportion of the power will generally be reflected at the junction. This reflection is avoided by the special measures taken here (impedance transformer and short-circuited line segment), at least at the design frequency. At this frequency, there are three admittances in parallel at

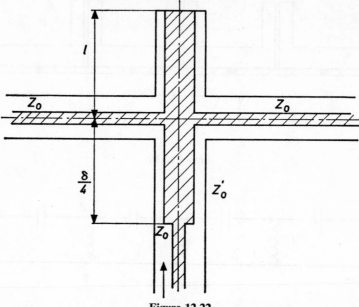

Figure 12.22

the junction, Y_0, Y_0 and $\mathrm{j}Y_0' \cot 2\pi l/\lambda$. The last term is zero, since $l = \lambda/4$; the total admittance is thus $2Y_0$. If the impedance transformer is to transform this $2Y_0$ to the Y_0 of the input line, it must have a characteristic impedance of $Z_0' = (1/\sqrt{2}).Z_0$. We have neglected the parasitic impedances due to the changes in diameter, which can be corrected for by adjusting the lengths of the impedance transformer and the short-circuited line segment.

If now the electric length of the transformer and the short-circuited line segment differ from the design values—for example, because the frequency differs from the nominal—the circuit will still function adequately over quite a wide frequency range, because both components will be either too long or too short. This may be seen as follows.

The input admittance of the short-circuited line for $\lambda \neq \lambda_0$, where λ_0 is the nominal frequency, is

$$-\mathrm{j}Y_0' \cot \frac{2\pi l}{\lambda} = -\mathrm{j}Y_0' \cot \left(\frac{\pi}{2}\frac{1}{1+\delta}\right)$$

where $\delta = \Delta\lambda/\lambda_0$ and $\lambda = \lambda_0 + \Delta\lambda$. The load on the quarter-wave transformer is thus

$$2Y_0 - \mathrm{j}Y_0' \cot \left(\frac{\pi/2}{1+\delta}\right)$$

so that the input admittance of this transformer, which is too long or too short in the same way as the short-circuited line, is given by

$$Y_i = Y_0' \frac{2Y_0 - jY_0' \cot\left(\dfrac{\pi/2}{1+\delta}\right) + jY_0' \tan\left(\dfrac{\pi/2}{1+\delta}\right)}{Y_0' + j\left[2Y_0 - jY_0' \cot\left(\dfrac{\pi/2}{1+\delta}\right)\right] \tan\left(\dfrac{\pi/2}{1+\delta}\right)}$$

$$= Y_0' \frac{2Y_0 + jY_0'\left\{ \tan\left(\dfrac{\pi/2}{1+\delta}\right) - 1 \middle/ \left[\tan\left(\dfrac{\pi/2}{1+\delta}\right)\right]\right\}}{2Y_0' + j2Y_0 \tan\left(\dfrac{\pi/2}{1+\delta}\right)}$$

Now if $\Delta\lambda$ is not too large, so that

$$\tan\left(\frac{\pi/2}{1+\delta}\right) \gg \frac{1}{\tan\left(\dfrac{\pi/2}{1+\delta}\right)}$$

then

$$Y_i \approx Y_0' \frac{2Y_0 + jY_0' \tan\left(\dfrac{\pi/2}{1+\delta}\right)}{2Y_0' + j2Y_0 \tan\left(\dfrac{\pi/2}{1+\delta}\right)} = \sqrt{2}Y_0 \frac{2Y_0 + j\sqrt{2}Y_0 \tan\left(\dfrac{\pi/2}{1+\delta}\right)}{2\sqrt{2}Y_0 + j2Y_0 \tan\left(\dfrac{\pi/2}{1+\delta}\right)} = Y_0$$

There is thus no noticeable reflection as yet.

A fixed power divider, which splits the original energy into two parts of different magnitude, and which is a modification of the divider we have just discussed, is shown in Fig. 12.23. We make the same assumptions as above, but now we make

$$Y_1 + Y_2 = 2Y_0$$

and

$$Y_1/Y_2 = P_1/P_2$$

with the aid of a number of quarter-wave impedance transformers; P_1 and P_2 are the powers into which the original power $P = P_1 + P_2$ is to be split.

The total admittance at the point of intersection of the lines for $\lambda = \lambda_0$ again is $2Y_0$, so that the input is again matched via the $\sqrt{2}.Y_0$ impedance transformer. Since the powers in the side-arms have a ratio equal to that of their input admittances, the condition that the energy should be split up into two unequal parts is also satisfied. It can be shown by a similar argument that the divider works well over a fairly wide band. For $\lambda = \lambda_0$,

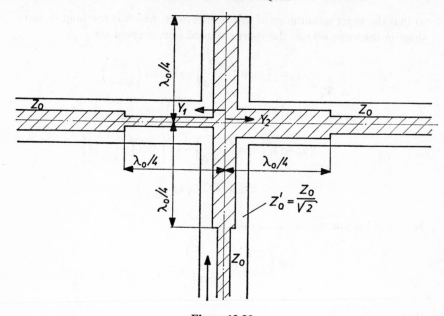

Figure 12.23

Y_1 and Y_2 are real; when $\lambda \neq \lambda_0$, the real parts scarcely differ in the first instance, while the imaginary parts have opposite signs. The bandwidth is less than in the previous case, because of the asymmetry.

A similar but slightly more complicated arrangement can be used to provide a variable power divider with continual matching of the input (for $\lambda = \lambda_0$). The power in either one of the side-arms can vary from almost zero to nearly the whole input power. The finite resistances of the short-circuit plungers used mean that the maximum 'damping' of the signal in a side-arm compared with that in the main line is about 45 dB. The circuit is shown in Fig. 12.24. All lines have the same characteristic impedance. A number of quarter-wave line segments are used, as indicated in the figure. The system is designed so that $l_D = l_C - \lambda/4$ when the plungers are moved in the stubs. We shall now discuss how the input power is divided between the outputs A and B, for various particular values of l_C.

1. $l_C = \lambda/2$, i.e. $l_D = \lambda/4$

In this case, stub C gives rise to a short-circuit at junction α, so that no power passes to output A. This short-circuit at α is a distance $\lambda/4$ from junction γ, so that it looks as if the whole branch A is not present at all. This branch presents an impedance $Z = \infty$ as seen from γ. If stub D

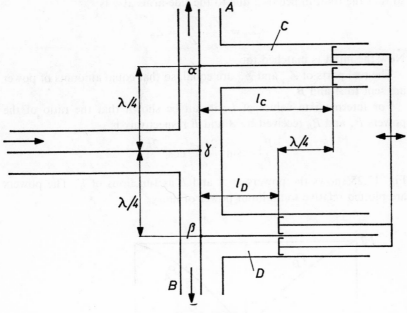

Figure 12.24

causes no reflection, all the energy goes to output B. As $l_D = \lambda/4$, stub D has an open impedance as seen from β, so that it looks as if D were not there either.

2. $l_C = 3\lambda/4$, *i.e.* $l_D = \lambda/2$

The same as above, except that now all the energy goes to A.

3. $l_C = 5\lambda/8$, *i.e.* $l_D = 3\lambda/8$

Stub D is now capacitive as seen from β, and stub C is inductive as seen from α. These reactances are in parallel with the matched loads of outputs B and A respectively. Putting the lengths chosen for C and D into the expression for the input impedance of a short-circuited line, we find that $Z_D = -jZ_0$ and $Z_C = +jZ_0$ as seen from β and α respectively. It follows that

$$Z_\alpha = jZ_0/(1+j) \quad \text{and} \quad Z_\beta = -jZ_0/(1-j)$$

The quarter-wave line segments from α and β to γ respectively transform these impedances to

$$Z_{\gamma_1} = \frac{Z_0^2}{Z_\alpha} = Z_0(1-j) \quad \text{and} \quad Z_{\gamma_2} = \frac{Z_0^2}{Z_\beta} = Z_0(1+j)$$

so that the total impedance due to the side-arms at γ is

$$Z_\gamma = \frac{Z_{\gamma_1} Z_{\gamma_2}}{Z_{\gamma_1} + Z_{\gamma_2}} = Z_0$$

Now the input is matched too.

The real parts of Z_{γ_1} and Z_{γ_2} are equal, so that equal amounts of power are sent to A and B.

For intermediate values of l_C, it can be shown that the ratio of the powers P_A and P_B received by A and B respectively is

$$\frac{P_A}{P_B} = \sin^2 \frac{2\pi l_C}{\lambda} \bigg/ \cos^2 \frac{2\pi l_C}{\lambda}$$

Fig. 12.25 shows the powers at A and B as functions of l_C. The powers are plotted relative to an input power of unity.

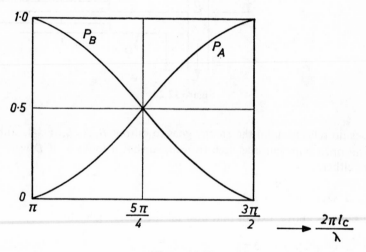

Figure 12.25

12.9 Measurement techniques

Without going into detail about the problems involved, we shall now describe a technique for measuring the attenuation of a coaxial component. The minimum equipment needed for this purpose is a signal generator for the desired frequency, a detector which is sensitive in the wavelength range in question and a meter such as a high-ohmic voltmeter for indicating the measured voltage.

One method of measuring the attenuation is the substitution method (Fig. 12.26). The input power is kept constant. Firstly the detected voltage

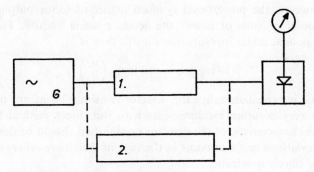

1. component under investigation
2. calibrated variable attenuator

Figure 12.26

is measured when the component in question is placed between the source and the detector. The component is then replaced by a calibrated variable attenuator, which is adjusted until the meter gives the same deflection as before. The loss of the component in question can now be read off from the calibrated attenuator.

A more direct measuring method is indicated in Fig. 12.27. The meter reading with the component under investigation switched on is n_1, and that with the component switched off is n_2. Let $n = n_1/n_2$ be the ratio of these readings.

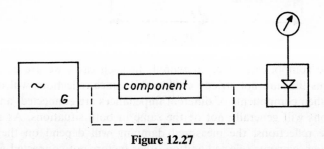

Figure 12.27

The way in which the attenuation can be calculated from n depends on the response of the detector to the incident power. At low powers (detector output-voltage less than a few tens of millivolts), the detector reacts quadratically, i.e. the output voltage is proportional to the square of the input voltage. In this case, the attenuation sought (in dB) is

$$10 \log_{10} n = 10 \log_{10} \frac{n_2}{n_1}$$

If, however, the power level is much higher (detector output voltage of the order of volts or more), the detector reacts linearly. The loss of the component under investigation (in dB) then is

$$10 \log_{10} n^2 = 10 \log_{10} \left(\frac{n_2}{n_1}\right)^2$$

In a wide intermediate range the detector is neither linear nor quadratic, and for very accurate measurements with this direct method the exact response characteristic of the detector crystal used should be determined. Small deviations can also occur in the regions which are otherwise purely linear or purely quadratic.

The advantage of the substitution method illustrated above is that the measurement can be made quite independent of the diode characteristic, since the meter reading is always adjusted to the same value.

Both of the above-mentioned methods can, however, give rise to other errors unless special measures are taken. Although the calibrated attenuator will generally be made with input and output impedances equal to the characteristic impedance of the transmission line, this is by no means always the case with the power sources, the detectors and in particular

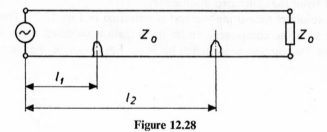

Figure 12.28

with the components to be measured. In such cases we are comparing situations that are not strictly comparable, since reflections will possibly occur when components of different impedances are connected, and these reflections will generally not be the same in both situations. As a result of these reflections, the measured damping will depend on the length of the line segments selected and on other factors not connected with the actual damping. This can cause very large relative errors when small dampings are measured; even reflections caused by connectors which are not perfectly matched can give rise to such large errors.

To illustrate the influence of the line length on errors in the measured reflection coefficient due to the presence of various disturbances, we shall consider the situation shown in Fig. 12.28. In a line with a matched load, disturbances are present at the points l_1 and l_2, which give rise to local reflections $\rho_1 e^{j\varphi_1}$ and $\rho_2 e^{j\varphi_2}$ respectively. Let $l_2 - l_1 = l$.

The total reflection at l_1 is found by adding the complex reflection coefficients there.

$$\rho\, e^{j\varphi} = \rho_1\, e^{j\varphi_1} + \rho_2\, e^{j\varphi_2}\, e^{j2\pi l/\lambda} = \rho_1\, e^{j\varphi_1} + \rho_2\, e^{j(\varphi_2 + 2\pi l/\lambda)}$$

The values of ρ and φ clearly depend on l. For example, if

$$\varphi_2 + \frac{2\pi l}{\lambda} = \varphi_1 + k.2\pi$$

where k is an integer, then the resultant reflection coefficient is maximum, and if $\varphi_2 + 2\pi l/\lambda = \varphi_1 + (2k+1)\pi$, then ρ will be minimum under these conditions.

In such cases, therefore, the value of the attenuation measurements is dubious. If possible, all the components used should be matched to the line. However, this is not always possible, and in such cases we must make use of 'directional isolators' to mask the effect. Directional isolators are components which make use of ferrites in such a way that the transmission properties depend strongly on the direction of propagation of the wave passing through them (they are thus non-reciprocal elements). In a limited frequency range, the attenuation is small in one direction and large in the other. The various parts of the circuit, with their particular sources of reflections, can now be decoupled by placing such isolators between them. The input and output impedances of the isolators are matched to the transmission line used.

It is also possible, though less elegant, to use fixed attenuators placed

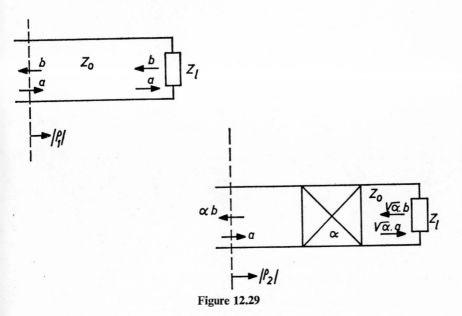

Figure 12.29

on both sides of the components to be measured. The reflection coefficient due to a mismatching further back in the circuit is then reduced, as shown in Fig. 12.29. The attenuator causes an energy attenuation of $1/\alpha$. The waves shown in the figure have arbitrary amplitudes of a and b, and the line does not have a matched load. Without the attenuator, the reflection coefficient would be

$$|\rho_1| = \frac{Z_l - Z_0}{Z_l + Z_0} = \frac{b}{a}$$

With the attenuator, it is

$$|\rho_2| = \alpha \frac{Z_l - Z_0}{Z_l + Z_0} = \alpha \frac{b}{a}$$

If, for example, $Z_l = 1 \cdot 5 Z_0$, then $|\rho_1| = 0 \cdot 5/2 \cdot 5 = 0 \cdot 2$, $|\rho_1|^2 = 0 \cdot 04$ and $|\rho_2| = 0 \cdot 2\alpha$, $|\rho_2|^2 = 0 \cdot 04\alpha^2$. With a tenfold attenuation, therefore, $|\rho_2|^2 = 0 \cdot 0004$.

In most cases, this method reduces any reflections that may occur to such a level that reasonable measurements can be made. The attenuators used must, of course, be matched to the line, especially in the substitution method.

WAVEGUIDES

We continue our discussion of wave-bearing systems with hollow, closed waveguides. Only these are normally given the name 'waveguide', the general term 'transmission line' being used to denote all the types of guide we have described up till now. We shall deal with the transmission properties of waveguides as well as the many different modes that can occur in them and the various components that are commonly used in this type of transmission.

13 WAVE PROPAGATION IN RECTANGULAR WAVEGUIDES

13.1 **Introduction**

We saw in Chapter 8 how the extension of plane waves could be restricted in one direction by placing two parallel conducting planes at right angles to the electric field. It was also shown that the plane waves obtained in this way still had an infinite energy content. In this section on waveguides, we shall see how the wave can be further confined by placing two more conducting planes at right angles to the first pair, thus giving a closed rectangular pipe. The wave is now confined within the pipe, and will propagate in its direction.

The situation obtained in this way is illustrated in Fig. 13.1. The electric field is perpendicular to the top and bottom of the waveguide, so that the condition $E_t = 0$ at the surface of a perfect conductor is

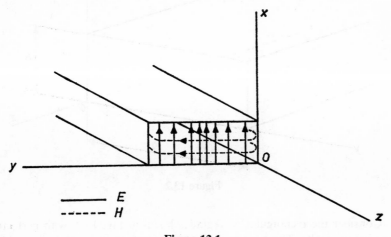

Figure 13.1

satisfied. The situation is quite different at the sides of the pipe, because the original field in a given xOy plane was independent of y (and x). Here, however, the electric field tangential to these conductors must be

143

zero, which can only happen if the electric field strength falls off from the middle of the pipe until it just reaches zero at the walls. But if the electric field strength varies in the y direction, this means that apart from $\partial E_x/\partial z \neq 0$ we now also have $\partial E_x/\partial y \neq 0$; and it necessarily follows from Maxwell's equations that $\partial B_z/\partial t \neq 0$ and $\partial B_y/\partial t \neq 0$. The magnetic field, therefore, also has components which vary with time in the y and z directions: $H_y \neq 0$ and $H_z \neq 0$. The field configuration is no longer that of a plane wave; it is no longer a transverse field, because the magnetic field now has a z component. A possible configuration of the lines of force is shown in Fig. 13.1. The E lines are all in the x direction. The H lines, which must be closed to satisfy the condition $\int_s B_n \, dS = 0$, curve round at the edges as the electric field strength decreases, and meet there.

After this qualitative introduction, we must try to solve Maxwell's equations for the boundary conditions found in the waveguide. We have shown that it is likely that waves can be propagated inside such a hollow pipe as well as in free space if the field configuration is suitably modified. The precise picture can be obtained by solving Maxwell's equations for the space in question.

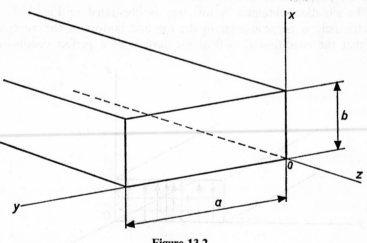

Figure 13.2

Consider the rectangular waveguide shown in Fig. 13.2, with perfectly conducting walls. There is an electromagnetic field inside the waveguide, but not outside. A possible wave in the guide will generally have components in all directions. If we can determine the electric field, for example, we can derive the magnetic field from it with the aid of Maxwell's second law. We start, therefore, by finding a solution for the electric field, without at present placing any general restrictions on time and place dependence. There are nevertheless certain conditions which the electric field (with components E_x, E_y and E_z) must satisfy.

Firstly, there are no free charges in the waveguide, $\rho = 0$, and it follows from equation (3.23) that

$$\frac{\partial E_x}{\partial x} + \frac{\partial E_y}{\partial y} + \frac{\partial E_z}{\partial z} = 0 \qquad (13.1)$$

Secondly, the electric field must satisfy Maxwell's equations, and thus the wave equations derived from Maxwell's equations with the aid of equation (13.1), for the three components of the electric field

$$\frac{\partial^2 E_x}{\partial x^2} + \frac{\partial^2 E_x}{\partial y^2} + \frac{\partial^2 E_x}{\partial z^2} = \varepsilon\mu \frac{\partial^2 E_x}{\partial t^2} \qquad (13.2)$$

$$\frac{\partial^2 E_y}{\partial x^2} + \frac{\partial^2 E_y}{\partial y^2} + \frac{\partial^2 E_y}{\partial z^2} = \varepsilon\mu \frac{\partial^2 E_y}{\partial t^2} \qquad (13.3)$$

$$\frac{\partial^2 E_z}{\partial x^2} + \frac{\partial^2 E_z}{\partial y^2} + \frac{\partial^2 E_z}{\partial z^2} = \varepsilon\mu \frac{\partial^2 E_z}{\partial t^2} \qquad (13.4)$$

These equations are derived as follows: we differentiate equation (3.18) with respect to z and equation (3.16) with respect to y, and add. The result is

$$-\frac{\partial}{\partial x}\left(\frac{\partial E_y}{\partial y} + \frac{\partial E_z}{\partial z}\right) + \frac{\partial^2 E_x}{\partial y^2} + \frac{\partial^2 E_x}{\partial z^2} = \mu\left(\frac{\partial^2 H_z}{\partial y\,\partial t} - \frac{\partial^2 H_y}{\partial z\,\partial t}\right)$$

Differentiation of equation (3.20) with respect to t (for $J_x = 0$) and substitution in the above expression gives equation (13.2), with the aid of equation (13.1). The wave equations for E_y and E_z can be obtained by cyclic permutation, or directly from Maxwell's equations by a procedure similar to that given above. This is rather a clumsy way of deriving the wave equations—the use of vector notation would have been more elegant.

We assume in the above argument that no conduction currents flow in the medium filling the waveguide, i.e. $\gamma = 0$ for this medium (see equation 3.22).

Thirdly, the field must satisfy the boundary condition that the tangential electric field is zero at the conducting walls of the waveguide

$$\begin{aligned} E_y = E_z = 0 \quad \text{for } x = 0 \text{ and } x = b \\ E_x = E_z = 0 \quad \text{for } y = 0 \text{ and } y = a \end{aligned} \qquad (13.5)$$

Now we are not interested in the most general solution of equations (13.1)–(13.5); we only wish to know if there are solutions which represent travelling waves in the z direction. Moreover, we can restrict our attention to waves which have a single harmonic time dependence, without loss of generality. This means that we look for solutions which can be written as follows in the complex notation

$$E_x(x, y, z, t) = G_x(x, y) e^{j\omega t - \Gamma z}$$
$$E_y(x, y, z, t) = G_y(x, y) e^{j\omega t - \Gamma z} \qquad (13.6)$$
$$E_z(x, y, z, t) = G_z(x, y) e^{j\omega t - \Gamma z}$$

Substitution in equations (13.1)–(13.4) gives the following equations for G_x, G_y and G_z

$$\frac{\partial G_x}{\partial x} + \frac{\partial G_y}{\partial y} - \Gamma G_z = 0 \qquad (13.7)$$

$$\frac{\partial^2 G_x}{\partial x^2} + \frac{\partial^2 G_x}{\partial y^2} = -(\Gamma^2 + \omega^2 \varepsilon \mu) G_x \qquad (13.8)$$

$$\frac{\partial^2 G_y}{\partial x^2} + \frac{\partial^2 G_y}{\partial y^2} = -(\Gamma^2 + \omega^2 \varepsilon \mu) G_y \qquad (13.9)$$

$$\frac{\partial^2 G_z}{\partial x^2} + \frac{\partial^2 G_z}{\partial y^2} = -(\Gamma^2 + \omega^2 \varepsilon \mu) G_z \qquad (13.10)$$

These have to be satisfied together with equation (13.5).

Equations (13.8)–(13.10) are identical, so we need to solve only one to arrive at the general solution.

We have so far been taking in account all three components of the electric and magnetic field. However, in view of the linearity of Maxwell's equations, we can always resolve such general waves into two kinds of wave, each of which has a simpler structure. One kind, called the TE or H configuration, has no longitudinal E component (TE stands for transverse electric), while the other, the TM or E configuration (transverse magnetic) has no longitudinal magnetic component. Making use of these, we shall solve the above equations for the two types of mode separately. Later, if we so wish, we can find a more general expression for the waves in the z direction by means of a linear combination of these solutions.

13.2 TE or H waves

There is no longitudinal component of the electric field

$$E_z = 0 \qquad (13.11)$$

This means that part of the boundary conditions is automatically satisfied. E_x and E_y (or G_x and G_y), as found by solving equations (13.8) and (13.9), must also satisfy equations (13.5) and (13.7). We shall now consider equation (13.8); that for G_y is identical.

This partial differential equation for $G_x(x, y)$ can be solved by separation of the variables to give two ordinary differential equations. We assume for this purpose that the unknown function G_x of x and y can be written as a product of a function of x alone and a function of y alone

$$G_x(x, y) = X(x)Y(y) \tag{13.12}$$

We shall now see whether there are any possible solutions of this type. Substituting this equation in (13.8), and dividing both sides by XY, we find

$$\frac{1}{X}\frac{d^2X}{dx^2} + \frac{1}{Y}\frac{d^2Y}{dy^2} = -(\Gamma^2 + \omega^2\varepsilon\mu) \tag{13.13}$$

The terms contain only ordinary, and not partial, derivatives. Moreover, the variables x and y are now separated, i.e. there is no term in the equation which contains both x and y. Since the first term is an arbitrary function of x alone, the second of y alone and the right-hand side is constant, both the first and second terms must be constant if equation (13.13) is to be satisfied for arbitrary values of x and y. Let us call these constants $-M_1^2$ and $-N_1^2$ respectively. We then have

$$\frac{1}{X}\frac{d^2X}{dx^2} = -M_1^2 \tag{13.14}$$

$$\frac{1}{Y}\frac{d^2Y}{dy^2} = -N_1^2 \tag{13.15}$$

where

$$M_1^2 + N_1^2 = \Gamma^2 + \omega^2\varepsilon\mu \tag{13.16}$$

The equations for $X(x)$ and $Y(y)$ also have the same form. Let us restrict our attention to equation (13.14); we want to find the solution to

$$\frac{d^2X}{dx^2} + M_1^2X = 0 \tag{13.17}$$

where $X = X(x)$.

The solution can be written in the complex notation or in terms of trigonometric functions. The latter is more convenient for the present problem

$$X(x) = A_1 \cos M_1 x + B_1 \sin M_1 x \tag{13.18}$$

where A_1 and B_1 are integration constants (arbitrary for the moment). It can be verified by substitution in equation (13.17) that (13.18) is indeed a solution.

Similarly, we find for $Y(y)$, from equation (13.15)

$$Y(y) = C_1 \cos N_1 y + D_1 \sin N_1 y \tag{13.19}$$

where C_1 and D_1 are arbitrary constants.

With the aid of equation (13.12), we thus find that

$$G_x(x, y) = (A_1 \cos M_1 x + B_1 \sin M_1 x)(C_1 \cos N_1 y + D_1 \sin N_1 y) \quad (13.20)$$

where $M_1^2 + N_1^2 = \Gamma^2 + \omega^2 \varepsilon \mu$.

We thus have the solution of equation (13.8) and it can be shown in the same way that the solution of equation (13.9) is

$$G_y(x, y) = (A_2 \cos M_2 x + B_2 \sin M_2 x)(C_2 \cos N_2 y + D_2 \sin N_2 y) \quad (13.21)$$

where $M_2^2 + N_2^2 = \Gamma^2 + \omega^2 \varepsilon \mu$.

These solutions must also satisfy equation (13.7) and the boundary conditions of equation (13.5). In the first of these equations, of course, $G_z(x, y) = 0$ for TE waves. With the aid of these equations, a number of restrictions can be imposed on the arbitrary constants which have been introduced above.

Since E_x must be zero, and hence $G_x = 0$ for $y = 0$, it follows from equation (13.20) that

$$C_1 = 0 \quad (13.22)$$

Similarly, the condition $E_y = 0$ and $G_y = 0$ for $x = 0$ gives, from equation (13.21)

$$A_2 = 0 \quad (13.23)$$

Since $D_1 \neq 0$—we would otherwise have $G_x \equiv 0$, which restriction is unnecessary in view of equation (13.24)—the condition $E_x = 0$ and hence $G_x = 0$ for $y = a$ and arbitrary x gives $\sin N_1 a = 0$, or

$$N_1 = \frac{n\pi}{a} \quad (13.24)$$

where $n = 0, 1, 2, 3, \ldots$

Similarly, the condition $E_y = 0$ and hence $G_y = 0$ for $x = b$ and arbitrary y gives $\sin M_2 b = 0$, or

$$M_2 = \frac{m\pi}{b} \quad (13.25)$$

where $m = 0, 1, 2, 3, \ldots$, since we assume $B_2 \neq 0$, to avoid having $E_y \equiv 0$.

The boundary condition of equation (13.5) is now satisfied.

Substituting the values of A_2, C_1, N_1, M_2 and G_z in the remaining condition, which is formulated in equation (13.7), we find

$$-A_1 D_1 M_1 \sin \frac{n\pi}{a} y \sin M_1 x - B_2 N_2 C_2 \sin N_2 y \sin \frac{m\pi}{b} x +$$

$$+B_1 D_1 M_1 \sin \frac{n\pi}{a} y \cos M_1 x + B_2 N_2 D_2 \sin \frac{m\pi}{b} x \cos N_2 y = 0 \quad (13.26)$$

This condition must be satisfied for all x and y within the waveguide and on its walls. We shall first consider a few simple restricted ranges within which this condition must be satisfied, and thus build up a general picture gradually. For $y = 0$ and $x < b$, the condition becomes

$$B_2 N_2 D_2 \sin \frac{m\pi}{b} x = 0$$

This can be satisfied by choosing $B_2 = 0$ or $D_2 = 0$. However, if we choose $B_2 = 0$, we should have E_y equal to zero. We therefore choose the less stringent requirement

$$D_2 = 0 \qquad (13.27)$$

Similarly, for $x = 0$ and $y \leqslant a$, and $D_1 \neq 0$ (since otherwise E_z would be zero), we find

$$B_1 = 0 \qquad (13.28)$$

Equation (13.26) is thus reduced to

$$A_1 D_1 M_1 \sin \frac{n\pi}{a} y \sin M_1 x + B_2 C_2 N_2 \sin N_2 y \sin \frac{m\pi}{b} x = 0 \quad (13.29)$$

for all $x \leqslant b$ and $y \leqslant a$.

It is preferable to choose all the remaining coefficients (A_1, D-, B_2 and C_2) non-zero in order to avoid either E_z or E_y being zero. This means that for $x = b$, $y \leqslant a$ we must have $\sin M_1 b = 0$, and for $y = a$, $x \leqslant b$ we must have $\sin N_2 a = 0$.

It follows that $M_1 = l\pi/b$, where $l = 0, 1, 2, \ldots$, and $N_2 = k\pi/a$, where $k = 0, 1, 2, \ldots$, However, we should take $l = m$ and $k = n$, since it will otherwise not be possible to satisfy equation (13.29) for $x < b$ and $y < a$. Hence

$$M_1 = M_2$$

and

$$N_1 = N_2 \qquad (13.30)$$

Substitution in equation (13.29) gives the final requirement

$$\frac{m\pi}{b} A_1 D_1 + \frac{n\pi}{a} B_2 C_2 = 0 \qquad (13.31)$$

Substitution of these results into equation (13.6) gives the equations for the electric field of TE waves in the waveguide

$$E_x = A_0 \frac{n}{a} \cos \frac{m\pi}{b} x \sin \frac{n\pi}{a} y \, e^{j\omega t - \Gamma z} \qquad (13.32)$$

$$E_y = -A_0 \frac{m}{b} \sin \frac{m\pi}{b} x \cos \frac{n\pi}{a} y \, e^{j\omega t - \Gamma z} \qquad (13.33)$$

$$E_z = 0 \qquad (13.34)$$

where

$$\Gamma^2 = \left(\frac{n\pi}{a}\right)^2 + \left(\frac{m\pi}{b}\right)^2 - \omega^2 \varepsilon \mu \qquad (13.35)$$

and

$$\frac{a}{n} A_1 D_1 = A_0 \qquad (13.36)$$

The corresponding magnetic field can be found from these equations for the electric field together with Maxwell's equations

$$H_x = -\frac{\Gamma}{j\omega\mu} E_y = A_0 \frac{m}{b} \frac{\Gamma}{j\omega\mu} \sin \frac{m\pi}{b} x \cos \frac{n\pi}{a} y \, e^{j\omega t - \Gamma z} \qquad (13.37)$$

M.T.

6

$$H_y = \frac{\Gamma}{j\omega\mu} E_x = A_0 \frac{n}{a} \frac{\Gamma}{j\omega\mu} \cos \frac{m\pi}{b} x \sin \frac{n\pi}{a} y \, e^{j\omega t - \Gamma z} \qquad (13.38)$$

$$H_z = \frac{A_0}{j\omega\mu\pi} \left\{ \left(\frac{n\pi}{a}\right)^2 + \left(\frac{m\pi}{b}\right)^2 \right\} \cos \frac{m\pi}{b} x \cos \frac{n\pi}{a} y \, e^{j\omega t - \Gamma z} \quad (13.39)$$

These last six equations thus represent the form of the electromagnetic field which can be produced in a waveguide, i.e. travelling waves in the z direction, varying harmonically with time, while in a plane at right angles to the direction of propagation ($z = $ constant) the field pattern is described by the above equations. The arbitrary constant A_0 is of course an amplitude factor.

However, the field derived above is only a part of the answer since it applies only to TE waves. Together with the TM waves we shall be discussing below, these field patterns allow a complete description of waves travelling in the z direction in waveguides. We shall not discuss attenuation until we start considering restrictions on the dimensions of the guide.

It should be noted that the six field equations given above still contain the quantities m and n, each of which can be chosen equal to zero or any integer, independently of each other. In other words, the equations should really be regarded as a double summation over all possible values of m and n if we want to represent a complete field pattern. For each combination of values of m and n, these six equations fix a certain field pattern. Now the field pattern corresponding to this set (m, n) is called a mode, in the present case a $TE_{m,n}$ mode. Each resultant component of the field can thus be built up of a linear combination of these modes. For example, the TE_{01} mode is the field pattern obtained by substituting $m = 0$ and $n = 1$ in our equations. The electric field of the TE_{01} mode can thus be written

$$E_x = \frac{A_0}{a} \sin \frac{\pi}{a} y \, e^{j\omega t - \Gamma z}$$
$$E_y = E_z = 0 \qquad\qquad (13.40)$$

As we shall see below, a corresponding set of TM modes exists. However, we shall also see that it is not always possible for all waves to propagate themselves in travelling-wave form in a guide of given transverse dimensions.

Note: It is not always possible or desirable to build up the general field pattern from elementary modes, as described above. For example, in situations where part of the waveguide or part of the space between the inner and outer conductor in a coaxial line is filled with a medium different from that in the rest of the space, it is often undesirable to split the field in this way. In general, however, and certainly within the framework of the present treatment, this method may be used to advantage.

13.3 TM or E waves

By definition, the longitudinal component of the transverse magnetic waves we are now about to consider is $H_z \equiv 0$. By the use of Maxwell's equations, this condition can also be expressed as a requirement for the electric field

$$\frac{\partial E_x}{\partial y} \equiv \frac{\partial E_y}{\partial x} \qquad (13.41)$$

We must now solve the wave equations (13.8)–(13.10) in much the same way as we did above, taking into account the boundary conditions in equation (13.5) and the subsidiary requirements of equations (13.7) and (13.41).

The general solution of the wave equations is completely analogous to that for the TE case. We find

$$G_x = (A_1 \cos M_1 x + B_1 \sin M_1 x)(C_1 \cos N_1 y + D_1 \sin N_1 y)$$
$$G_y = (A_2 \cos M_2 x + B_2 \sin M_2 x)(C_2 \cos N_2 y + D_2 \sin N_2 y) \qquad (13.42)$$
$$G_z = (A_3 \cos M_3 x + B_3 \sin M_3 x)(C_3 \cos N_3 y + D_3 \sin N_3 y)$$

As above, it follows from the boundary conditions that
for $E_z = 0$ and hence $G_z = 0$ at $x = 0$: $A_3 = 0$ and $B_3 \neq 0$ since otherwise $E_z \equiv 0$

$E_z = 0$	$G_z = 0$	$y = 0$:	$C_3 = 0$	$D_3 \neq 0$	$E_z \equiv 0$
$E_y = 0$	$G_y = 0$	$x = 0$:	$A_2 = 0$	$B_2 \neq 0$	$E_y \equiv 0$
$E_x = 0$	$G_x = 0$	$y = 0$:	$C_1 = 0$	$D_1 \neq 0$	$E_x \equiv 0$
$E_z = 0$	$G_z = 0$	$x = b$:	$M_3 = m\pi/b$		
$E_z = 0$	$G_z = 0$	$y = a$:	$N_3 = n\pi/a$		
$E_y = 0$	$G_y = 0$	$x = b$:	$M_2 = m'\pi/b$		
$E_x = 0$	$G_x = 0$	$y = a$:	$N_1 = n'\pi/a$		(13.43)

The condition (13.41) can be written, with the aid of equation (13.42) and the above restrictions

$$\frac{m'\pi}{b} B_2 C_2 \cos \frac{m'\pi}{b} x \left[\cos N_2 y + \frac{D_2}{C_2} \sin N_2 y \right]$$
$$= \frac{n'\pi}{a} A_1 D_1 \cos \frac{n'\pi}{a} y \left[\cos M_1 x + \frac{B_1}{A_1} \sin M_1 x \right] \qquad (13.44)$$

for all $x \leqslant b$ and $y \leqslant a$.

This condition can be satisfied by putting

$$B_1 = 0 \qquad D_2 = 0 \qquad N_2 = \frac{n'\pi}{a} \qquad M_1 = \frac{m'\pi}{b}$$

and

$$\frac{m'\pi}{b} B_2 C_2 = \frac{n'\pi}{a} A_1 D_1 \qquad (13.45)$$

We are still left with requirement (13.7) which, with the relations already found, may be written

$$\left(A_1 D_1 \frac{m'\pi}{b} + B_2 C_2 \frac{n'\pi}{a}\right) \sin \frac{m'\pi}{b} x \sin \frac{n'\pi}{a} y$$

$$= -\Gamma B_3 D_3 \sin \frac{m\pi}{b} x \sin \frac{n\pi}{a} y \quad (13.46)$$

for all $x < b$ and $y < a$.

These conditions can be satisfied by putting

$$m' = m \qquad n' = n \qquad (13.47)$$

and

$$\Gamma B_3 D_3 + A_1 D_1 \frac{m\pi}{b} + B_2 C_2 \frac{n\pi}{a} = 0$$

We now determine the magnetic field from the values found above for the electric field by using Maxwell's equations. Firstly, we may summarize the results found so far

$$G_x = A_1 D_1 \cos \frac{m\pi}{b} x \sin \frac{n\pi}{a} y$$

$$G_y = B_2 C_2 \sin \frac{m\pi}{b} x \cos \frac{n\pi}{a} y \qquad (13.48)$$

$$G_z = B_3 D_3 \sin \frac{m\pi}{b} x \sin \frac{n\pi}{a} y$$

The constants in (13.48) are further limited by the relations (13.45) and (13.47) found above. It is obvious that TM waves must also satisfy

$$\Gamma^2 = \left(\frac{m\pi}{b}\right)^2 + \left(\frac{n\pi}{a}\right)^2 - \omega^2 \varepsilon \mu \qquad (13.49)$$

The magnetic field thus becomes

$$H_x = -\frac{1}{j\omega\mu} \left(\frac{n\pi}{a} B_3 D_3 + \Gamma B_2 C_2\right) \sin \frac{m\pi}{b} x \cos \frac{n\pi}{a} y \, e^{j\omega t - \Gamma z}$$

$$H_y = \frac{1}{j\omega\mu} \left(\Gamma A_1 D_1 + \frac{m\pi}{b} B_3 D_3\right) \cos \frac{m\pi}{b} x \sin \frac{n\pi}{a} y \, e^{j\omega t - \Gamma z} \qquad (13.50)$$

$$H_z = 0$$

The two above-mentioned relations between the constants in these equations can now be used to eliminate some of the constants. For this purpose let

$$\frac{1}{j\omega\mu} \left(\Gamma A_1 D_1 + \frac{m\pi}{b} B_3 D_3\right) = \frac{m}{b} B_0 \qquad (13.51)$$

It follows from equations (13.45) and (13.47) that

$$\frac{1}{j\omega\mu}\left(\Gamma B_2 C_2 + \frac{n\pi}{a} B_3 D_3\right) = \frac{n}{a} B_0 \qquad (13.52)$$

The components of the magnetic field in TM waves can thus be written

$$H_x = -\frac{n}{a} B_0 \sin\frac{m\pi}{b}x\cos\frac{n\pi}{a}y\, e^{j\omega t - \Gamma z} \qquad (13.53)$$

$$H_y = \frac{m}{b} B_0 \cos\frac{m\pi}{b}x\sin\frac{n\pi}{a}y\, e^{j\omega t - \Gamma z} \qquad (13.54)$$

$$H_z = 0 \qquad (13.55)$$

Let us now also re-write the constants for the electric field by using the above restrictions and the nomenclature defined in equation (13.51). It follows from equation (13.45) that

$$B_2 C_2 = \frac{n\pi}{a}\frac{b}{m\pi} A_1 D_1 \qquad (13.56)$$

and from equation (13.5) that

$$B_3 D_3 = \frac{j\omega\mu}{\pi} B_0 - \Gamma A_1 D_1 \qquad (13.57)$$

Substitution of the last two equations in (13.47) gives

$$A_1 D_1 = \frac{\Gamma}{j\omega\varepsilon}\frac{m}{b} B_0 \qquad (13.58)$$

and hence, with equation (13.56)

$$B_2 C_2 = \frac{\Gamma}{j\omega\varepsilon}\frac{n}{a} B_0 \qquad (13.59)$$

and with equation (13.57)

$$B_3 D_3 = \frac{-B_0}{j\omega\varepsilon\pi}\left\{\left(\frac{m\pi}{b}\right)^2 + \left(\frac{n\pi}{a}\right)^2\right\} \qquad (13.60)$$

The final form for the electric field of the TM waves may thus be written

$$E_x = \frac{\Gamma}{j\omega\varepsilon}\frac{m}{b} B_0 \cos\frac{m\pi}{b}x\sin\frac{n\pi}{a}y\, e^{j\omega t - \Gamma z} = \frac{\Gamma}{j\omega\varepsilon} H_y \qquad (13.61)$$

$$E_y = \frac{\Gamma}{j\omega\varepsilon}\frac{n}{a} B_0 \sin\frac{m\pi}{b}x\cos\frac{n\pi}{a}y\, e^{j\omega t - \Gamma z} = \frac{-\Gamma}{j\omega\varepsilon} H_x \qquad (13.62)$$

$$E_z = \frac{-B_0}{j\omega\varepsilon\pi}\left\{\left(\frac{m\pi}{b}\right)^2 + \left(\frac{n\pi}{a}\right)^2\right\}\sin\frac{m\pi}{b}x\sin\frac{n\pi}{a}y\, e^{j\omega t - \Gamma z} \qquad (13.63)$$

The remarks made in the previous section concerning the solution for TE fields also apply to the TM field pattern. In general, therefore, an arbitrary waveguide can contain a complicated field configuration,

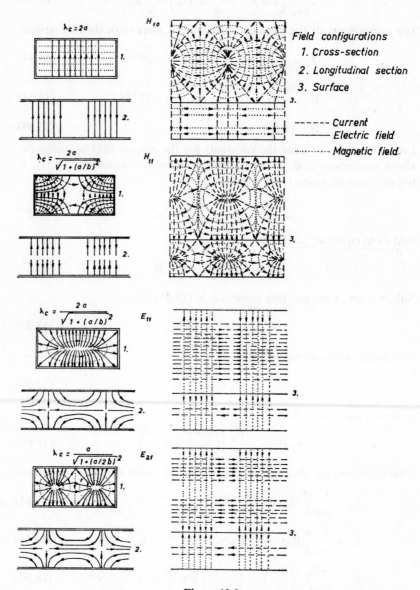

Figure 13.3

especially if we remember that the resultant field can be built up of a linear combination of TE and TM modes.

The field patterns of all the simplest modes (low values of m and n) are given in the literature. (See, for example, N. Marcuvitz, *Waveguide Handbook*, Radiation Laboratory Series, Vol. 10, pp. 59 and 63; also Fig. 13.3.) It should, however, be noted that $TM_{0,n}$ and $TM_{m,0}$ modes cannot exist since the electric field is equal to zero.

It is clearly a great advantage to know the precise form of the electric and magnetic field when designing components for use in the waveguide and similar purposes. It would, for example, be convenient if the waveguide could be made of such dimensions that not all TE and TM modes can exist or be propagated in it. For example, if only one mode were possible, the electromagnetic field would be much easier to deal with. We shall accordingly consider the propagation conditions for the various modes in the next chapter.

14 PROPERTIES OF RECTANGULAR WAVEGUIDES

14.1 Propagation constant and cut-off frequency

For considerations about the propagation of the TM and TE waves derived in Chapter 13 the common factor $\exp[j\omega t - \Gamma z]$ suffices here. The precise form of the electromagnetic field in a given cross-section where $z = $ constant is not relevant in this connection.

We have seen that the expression

$$\Gamma^2 = \left(\frac{m\pi}{b}\right)^2 + \left(\frac{n\pi}{a}\right)^2 - \omega^2 \varepsilon \mu$$

is valid for the propagation constant of both the $TE_{m,n}$ and the $TM_{m,n}$ modes.

If the frequency is low enough, we shall have $\Gamma^2 > 0$ and hence Γ real, while at high frequencies $\Gamma^2 < 0$ so that Γ is imaginary. Since we have not considered any dielectric or wall losses in our arguments so far, it follows that Γ cannot be complex.

Now what difference does it make to the propagation whether Γ is real or imaginary? If $\Gamma^2 < 0$, we may write

$$\Gamma = j\beta \tag{14.1}$$

where β is real. The electric field then has the form

$$E = G(x, y) \, e^{j(\omega t - \beta z)} \tag{14.2}$$

As we have already seen, this expression represents a travelling wave in the z_+ direction, which is not attenuated as long as the medium and the wall have no losses. The parameter β is given by

$$|\beta| = \sqrt{\omega^2 \varepsilon \mu - \left(\frac{m\pi}{b}\right)^2 - \left(\frac{n\pi}{a}\right)^2} \tag{14.3}$$

For each combination of values of m and n, β has its own value, which will in general differ from that for another (m, n) combination and from the velocity of light. Moreover, for a given mode β depends in a non-linear manner on the frequency. Wave propagation in a waveguide will thus always be dispersive (see Section 5.1).

If, however, $\Gamma^2 > 0$, so that Γ is real, the electric field will have the form

$$E = G(x, y) \, e^{j\omega t} \, e^{-\Gamma z} \tag{14.4}$$

When Γ is real, this does not represent a travelling wave since no energy is propagated in the z direction. The field strength of such a mode falls off exponentially in the z direction along the waveguide. This attenuation is not due to losses, since we have assumed these to be zero, but the phenomenon more closely resembles the attenuation of high frequencies in a low-pass filter, or the total reflection of light waves.

If the mode, the frequency and the dimensions of the guide are given, the decrease in the intensity of the field per unit length in the z direction can, of course, be determined, being Γ neper per unit length, or $8 \cdot 69 \Gamma$ dB per unit length. For the TE_{01} mode, this is

$$8 \cdot 69 \, \frac{2\pi}{2a} \sqrt{1 - \left(\frac{2a}{\lambda}\right)^2} = \frac{8 \cdot 69\pi}{a} \sqrt{1 - \left(\frac{2a}{\lambda}\right)^2} \quad \text{dB per unit length}$$

where λ is the wavelength in free space at the operating frequency.

Under what conditions do we find this effect, which limits the number of modes occurring in the rectangular waveguide to a few, only one, or even none? The answer to this question clearly can be found from the expression for the propagation constant. If the frequency of a certain mode (m and n known) in a given waveguide (a and b known) decreases from a high value, then at a certain frequency the value of Γ^2 will pass from negative (propagation possible) through zero to positive (propagation impossible). The limiting frequency at which this transition occurs is known as the *cut-off frequency* f_c, and the corresponding wavelength in free space, $\lambda_c = c/f_c$, the cut-off wavelength. The waveguide acts as a high-pass filter: high frequencies are transmitted without attenuation (in the absence of losses), while low frequencies cannot be propagated, the field strength decreasing exponentially along the waveguide. For a given guide we can find the cut-off frequency for an arbitrary mode simply by putting $\Gamma^2 = 0$, or

$$\omega_c^2 \varepsilon \mu = 4\pi^2 f_c^2 \varepsilon \mu = \left(\frac{m\pi}{b}\right)^2 + \left(\frac{n\pi}{a}\right)^2$$

or

$$f_c^2 = \frac{1}{\varepsilon \mu} \left\{ \left(\frac{m}{2b}\right)^2 + \left(\frac{n}{2a}\right)^2 \right\} \tag{14.5}$$

or

$$f_c^2 = \frac{c^2}{\varepsilon_r \mu_r} \left\{ \left(\frac{m}{2b}\right)^2 + \left(\frac{n}{2a}\right)^2 \right\} \qquad (14.6)$$

or, expressed in terms of wavelength

$$\lambda_c^2 = \frac{c^2}{f_c^2} = \varepsilon_r \mu_r . 1/\{(m/2b)^2 + (n/2a)^2\} \qquad (14.7)$$

It follows that for the TE_{01} mode ($m = 0$, $n = 1$) we have

$$\lambda_c = \sqrt{\varepsilon_r \mu_r} . 2a$$

and if the medium in the guide is air, this becomes

$$\lambda_c = 2a$$

In other words, with a given air-filled waveguide, waves with a free-space wavelength $\lambda > 2a$ cannot be propagated in the TE_{01} mode.

Similarly, the cut-off wavelength for the TE_{10} mode ($m = 1$, $n = 0$) in an air-filled waveguide is $\lambda_c = 2b$. Waves of a frequency f such that $\lambda > 2b$ cannot be propagated in the TE_{10} mode. We have assumed that a is the width of the waveguide, and b its height, and that $a > b$.

In general, then, a wave varying in time with a frequency f cannot be propagated in an air-filled waveguide if

$$\lambda = \frac{c}{f} > 1/\{(m/2b)^2 + (n/2a)^2\}^{\frac{1}{2}} \qquad (14.8)$$

As the frequency increases, more and more modes can be propagated freely through the guide (higher values of m and/or n are possible). In many cases it is desirable to choose the dimensions of the waveguide so that only one mode can be propagated at the desired operating frequency in order to facilitate control over the field distribution. How must a and b be chosen with respect to the wavelength to achieve this?

With TM modes, the greatest wavelength that can occur is the cut-off wavelength of the TM_{11} mode (as we have seen, $TM_{0,m}$ and $TM_{m,0}$ modes are impossible)

$$\lambda_{c\,TM_{1,1}} = 2ab/(a^2 + b^2)^{\frac{1}{2}}$$

(from equation 14.7).

With TE modes, the cut-off wavelength of the TE_{01} mode is

$$\lambda_{c\,TE_{0,1}} = 2a$$

that of the TE_{10} mode is

$$\lambda_{c\,TE_{1,0}} = 2b$$

and that of the TE_{11} mode is

$$\lambda_{c\,TE_{1,1}} = 2ab/(a^2 + b^2)^{\frac{1}{2}}$$

A wave of wavelength (in free space) greater than $2a$ cannot, therefore, be propagated. If the wavelength is between $2a$ and $2b$, propagation is only possible in the TE_{01} mode. For wavelengths between $2b$ and $2ab/(a^2+b^2)^{\frac{1}{2}}$ two forms are possible, namely the TE_{01} mode and the TE_{10} mode. Finally, if the wavelength is even less, at least the TE_{01}, TE_{10}, TE_{11} and TM_{11} modes can occur.

These considerations of cut-off frequency therefore show, among other things, that there is a certain frequency range in which only the TE_{01} mode will be propagated in a rectangular guide; this is called the *principal mode* of the waveguide. Of course, when the waveguide is used in this frequency range, bends in the guide or other discontinuities can give rise to higher modes locally, so that part of the energy originally present in the principal mode is converted into higher modes. Such higher modes cannot, however, be propagated freely along the waveguide, and a short distance after the discontinuity the intensity of these higher modes will be negligible, all the energy having reverted to the principal mode. This assumes, of course, that there are no dielectric or wall losses and that the discontinuity did not give rise to any reflection. (In the latter case, of course, even if no higher modes are produced part of the energy originally propagated in a given direction will be reflected in the opposite direction, and will probably be absorbed in the source or elsewhere.)

The wavelength (or frequency) range in which only the TE_{01} mode is propagated in a waveguide is given by

$$2b < \lambda < 2a$$

$$\frac{c}{2a} < f < \frac{c}{2b} \tag{14.9}$$

In practice, all waveguides are standardized so that the above relation is satisfied for the frequency range in which they are to be used. Standard waveguide dimensions for various wavelength ranges have been tabulated in many publications. Appendix 4 gives the dimensions of a number of waveguides in millimetres (converted from the original values, which were given in inches). Accepted names for various frequency ranges are also given, together with the numerical notation recommended by the CCIR (*Comité consultatif international de Radiocommunication*). The CCIR notation gives exact definitions of the various frequency ranges although this is not so with the names given in Appendix 4, which are nevertheless widely used in practice.

Higher frequencies can, of course, also be transmitted in a waveguide as defined by equation 14.9, but it then becomes difficult to predict which mode will occur. If higher modes do occur in a waveguide, the proportion of the various modes can vary from place to place, particularly when the waveguide contains discontinuities. It is then practically impossible to control the waves with components that have been designed for the lowest mode. Certain measures can be taken even in such cases to ensure that the waves are propagated mainly in the TE_{01} mode. This phenomenon is sometimes used to minimize the losses which occur in practice, particularly in waveguides of small transverse dimensions for use in the millimetre wavelength range. Higher modes which may occur in this case are suppressed with 'mode filters'. However, a detailed discussion of such 'oversized' waveguides would take us beyond the scope of this book.

Since we shall be dealing with the TE_{01} mode frequently from now on, the field equations for this mode, derived from the general equations (13.32)–(13.39), are given below

$$E_x = E_0 \sin \frac{\pi y}{a} e^{j\omega t - \Gamma z}$$

$$E_y = E_z = 0$$

$$H_x = 0$$

$$H_y = \frac{\Gamma}{j\omega\mu} E_z = \frac{\Gamma}{j\omega\mu} E_0 \sin \frac{\pi y}{a} e^{j\omega t - \Gamma z} \qquad (14.10)$$

$$H_z = \frac{\Gamma}{j\omega\mu} \frac{\pi}{a} \cos \frac{\pi y}{a} e^{j\omega t - \Gamma z}$$

where $\Gamma^2 = \pi^2/a^2 - \omega^2\varepsilon\mu$ and $\beta = 2\pi/\lambda_g$, and the cut-off wavelength $\lambda_c = 2a$.

The above equations also make use of the new symbol $E_0 = A_0/a$; the significance of λ_g will be discussed in the next section.

14.2 Phase velocity and group velocity; guide wavelength

We have already encountered the *phase velocity* and *group velocity* concepts. We shall now examine what they mean in the special context of waveguides.

We found above that the group velocity can be written

$$v_{gr} \approx 1 \Big/ \frac{d\beta}{d\omega}$$

With travelling waves in a rectangular waveguide, β is given by equation (14.3), from which it follows that

$$v_{gr} \approx \frac{1}{\sqrt{\varepsilon\mu}} \sqrt{1 - \frac{(m/b)^2 + (n/a)^2}{4\varepsilon\mu f^2}} = \frac{c}{\sqrt{\varepsilon_r \mu_r}} \sqrt{1 - \lambda^2/\lambda_c^2} \qquad (14.11)$$

If the guide is filled with air, both ε_r and μ_r are equal to unity; c is the velocity of light *in vacuo*. The group velocity for the TE_{01} mode in such a waveguide is given by

$$v_{gr\,TE_{01}} \approx c\sqrt{1 - (\lambda/2a)^2} \qquad (14.12)$$

The phase velocity is given by $v_{ph} = \omega/\beta$, so that for waveguides

$$v_{ph} = \frac{c}{\sqrt{\varepsilon_r \mu_r}} \frac{1}{\sqrt{1-(\lambda/\lambda_c)^2}} \qquad (14.13)$$

and for the TE_{01} mode in an air-filled waveguide

$$v_{ph\,TE_{01}} = \frac{c}{\sqrt{1-(\lambda/2a)^2}} \qquad (14.14)$$

In general, then, it follows from equations (14.11) and (14.13) that

$$v_{gr} \cdot v_{ph} = \frac{1}{\varepsilon\mu} = \frac{c^2}{\varepsilon_r \mu_r} \qquad (14.15)$$

or $v_{gr} \cdot v_{ph} = c^2$ in a waveguide with air as medium (see also Fig. 14.1). In all cases $v_{ph} > c$ and $v_{gr} < c$.

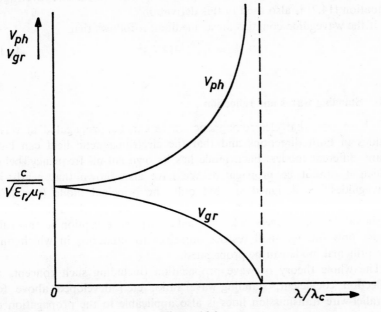

Figure 14.1

It can be seen from the expression for the phase velocity that v_{ph} is frequency-dependent, even for the TE_{01} mode in an air-filled waveguide. Unlike a coaxial line, a waveguide is always dispersive, even when it is filled with air. This dispersion increases the closer the cut-off wavelength is approached within the permissible wavelength range for a given waveguide. This could be another reason for using oversized waveguides, in which $\lambda \ll \lambda_c$, so that the waveguide is practically non-dispersive.

We have already mentioned that the wavelength of a wave of frequency f in a given medium is obtained by dividing the phase velocity in that medium by the frequency. However, the periodicity of the wave in a guide can also be determined with the aid of the expression (14.2). It follows from this that the wave is periodic in z, as $e^{-j\beta z}$. If βz increases by 2π, the wave has the same amplitude. The *guide wavelength* λ_g is thus given by $\beta\lambda_g = 2\pi$. The expression for the guide wavelength can thus be derived from that for β given in equation (14.3)

$$\lambda_g = \frac{2\pi}{\beta} = \frac{1}{\sqrt{\varepsilon_r\mu_r}} \cdot \frac{c}{f} \cdot \frac{1}{\sqrt{1-(\lambda/\lambda_c)^2}}$$

$$= \frac{1}{\sqrt{\varepsilon_r\mu_r}} \cdot \frac{\lambda}{\sqrt{1-(\lambda/\lambda_c)^2}} \tag{14.16}$$

where λ is the wavelength in free space and λ_c is the cut-off wavelength; equation (14.7) is also used in this derivation.

If the waveguide contains air as medium, it follows that

$$\lambda_g = \lambda/\sqrt{1-(\lambda/\lambda_c)^2} \tag{14.17}$$

14.3 Standing waves and reflection

We have seen that electromagnetic waves can be propagated in wave-guides in both directions and that the electromagnetic field can have many different modes. Each mode has its own cut-off frequency, below which it cannot be propagated. We have also shown that nearly all waveguides are designed so that only the principal mode (the TE_{01} mode) can be propagated in the frequency range for which the guide is designed, although oversized waveguides form an exception to this rule. From now on, we shall restrict ourselves to situations in which only the principal mode can be propagated.

The whole theory of wave propagation (including such concepts as reflection coefficient, standing-wave ratio, etc.) developed above for parallel-wire transmission lines is also applicable to the propagation of waves in waveguides, as we shall see below.

Consider a waveguide containing an obstacle which does not occupy the whole cross-section of the guide (for example, a diaphragm or a conducting wire joining the top and bottom of the guide). A travelling wave $A\,e^{-j\beta z}$ (Fig. 14.2) passes this obstacle from the left, where A is a function of the co-ordinates in a plane at right angles to the direction of propagation, $A = A(x, y)$. Near the obstacle, the fields of the principal mode, which are solutions of Maxwell's equations in the empty waveguide, will not satisfy the local boundary conditions (for example, that the

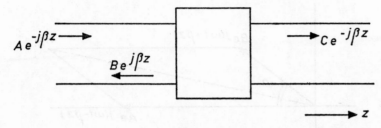

Figure 14.2

tangential component of the electric field strength should be zero at the surface of a perfect conductor). Higher modes will thus be needed locally to satisfy the new boundary conditions. However, these higher modes cannot be propagated in the waveguide and a short distance from the obstacle (for example, one guide wavelength) only the principal mode will remain. A more complicated field pattern is required, and present, only in the immediate neighbourhood of the obstacle. The overall effect of the obstacle on the wave is that the travelling wave established under the new steady-state conditions, $C\,e^{-j\beta z}$, has a smaller, complex amplitude than the incident wave: $|C| < |A|$. The difference in energy between the incident and the transmitted wave is accounted for by a wave $B\,e^{j\beta z}$ reflected by the obstacle, and thus travelling to the left. If, moreover, the obstacle is a high-loss one, some of the energy of the incident wave will also be dissipated in it, leading to the passage of current through the obstacle. For the moment we shall regard such obstacles, like the walls of the waveguide, as being lossless. The phase constant $\beta = 2\pi/\lambda_g$. As is usual in the representation of waves, the factor $e^{j\omega t}$ is omitted. In the complex notation used, the amplitudes A, B and C are generally complex.

At some distance, therefore, the effect of the obstacle is fully represented by the complex values of A, B and C. These do not, of course, tell us anything about the field configuration in the immediate neighbourhood of the obstacle, but this is not usually of interest. Without sacrificing generality, we can assume for the sake of simplicity that A is real. To the right of the obstacle, there is only one travelling wave, $C\,e^{j(\omega t - \beta z)}$. To the left, there are two, travelling in opposite directions, and their amplitudes A and B are proportional to the transverse electric field at any point in the transverse plane. Fig. 14.3 shows the situation at time $t = 0$ and position $z = 0$, by means of a diagram in the complex plane—here, for example, $B = |B|\,e^{j\phi}$. For a given transverse plane ($z = $ constant), both vectors rotate anticlockwise with angular velocity ω. The resultant thus has a constant modulus, but its phase is continually changing. In another transverse plane, the resultant field strength will have another constant modulus. If, however, we regard the situation at a given moment

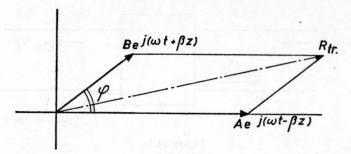

Figure 14.3

(t = constant) and move along the axis in the z_+ direction, then the vector $B\,e^{j(\omega t+\beta z)}$ rotates anticlockwise and the vector $A\,e^{j(\omega t-\beta z)}$ rotates clockwise. The resultant no longer has a constant modulus along the z axis, but exhibits maxima and minima (at $\varphi = 0°$ and $180°$ respectively).

As we have already found in similar cases, we thus have a travelling wave superimposed on a standing wave or, if $|B| = |A|$ and $|C| = 0$, a pure standing wave. The distance z_1 between the obstacle and the first minimum in the standing-wave pattern (Fig. 14.4) is given by

$$z_1 = \frac{\pi - \varphi}{2\beta} \tag{14.18}$$

It is, of course, possible to determine the resultant F of A and B in a more quantitative manner

$$F = A\,e^{j(\omega t-\beta z)} + B\,e^{j(\omega t+\beta z)} = e^{j\omega t}\left[A\,e^{-j\beta z} + B_1\,e^{j(\beta z+\varphi)}\right] \tag{14.19}$$

where $B_1 = |B|$

$$F.e^{-j\omega t} = (A\,e^{-j\beta z} - B_1\,e^{-j(\beta z+\varphi)}) + (B_1\,e^{-j(\beta z+\varphi)} + B_1\,e^{j(\beta z+\varphi)})$$

$$= D\,e^{-j\beta z+j\theta} + 2B_1\cos(\beta z+\theta)$$

where $D = |A - B_1\,e^{-j\phi}|$ and $\theta = \tan^{-1}(B_1\sin\varphi)/(A - B_1\cos\varphi)$. Hence

$$F = D.e^{j(\omega t-\beta z+\theta)} + 2B_1 e^{j\omega t}\cos(\beta z+\varphi) \tag{14.20}$$

i.e. a travelling wave of complex amplitude $D\,e^{j\theta}$ superimposed on a standing wave of amplitude $2B_1$.

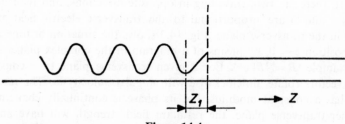

Figure 14.4

The standing-wave pattern produced in this wave can be measured with a standing-wave detector (a movable probe which scans the field in the guide in the longitudinal direction). Just as we found in the previous cases, we can now introduce the standing-wave ratio η, which is the ratio of the maximum and minimum field strengths in this pattern

$$\eta = \frac{|F|_{max}}{|F|_{min}} = \frac{|A|+|B|}{|A|-|B|} = \frac{A+B_1}{A-B_1} \tag{14.21}$$

Similarly, we can again define the reflection coefficient r as the ratio of the complex amplitudes of the reflected and incident fields

$$r = \frac{B\,e^{j\beta z}}{A\,e^{-j\beta z}} = \frac{B_1}{A}\,e^{j(2\beta z + \varphi)} \tag{14.22}$$

for an obstacle at the point z.

Now $B_1 \leqslant A$. As B_1/A varies between 0 and 1, $|r|$ also varies between 0 and 1. We always have $|r| \leqslant 1$; the phase of r varies with z. Just as we found before, η and r are related by the equations

$$\eta = \frac{A+B_1}{A-B_1} = \frac{1+B_1/A}{1-B_1/A} = \frac{1+|r|}{1-|r|} \tag{14.23}$$

$$|r| = (\eta-1)/(\eta+1) \tag{14.24}$$

At a minimum of the standing wave, the phase of r is π, and at a maximum the phase of r is zero.

Until now, we have assumed for the sake of simplicity that the wave $B\,e^{-j\beta z}$ reflected by the obstacle is absorbed by the source (matched source). It is, of course, possible that this may not happen and that the wave is partially reflected back and forth between the source and the obstacle. If such a multiple reflection is produced, the final result can be expressed in terms of a geometric series. If $r_1 = |r_1|\,e^{j\varphi_1}$ is the reflection coefficient of the obstacle and $r_2 = |r_2|\,e^{j\varphi_2}$ that of the source, then we have a series of waves travelling from left to right

$$A + Ar_1 r_2 + Ar_1^2 r_2^2 + \ldots = A/(1-r_1 r_2)$$

and from right to left

$$Ar_1 + Ar_1^2 r_2 + Ar_1^3 r_2^2 + \ldots = Ar_1/(1-r_1 r_2)$$

The only difference in the final situation is that the original incident wave appears to have an amplitude $A' = A/(1 - r_1 r_2)$ instead of A.

A number of power relations can now be expressed in terms of the reflection coefficient and the standing-wave ratio.

If the reflection coefficient is r, then the part of the incident wave which is reflected by the obstacle has an energy

$$P_r = rr^*P_{in} = |r|^2 P_{in} = \left(\frac{\eta-1}{\eta+1}\right)^2 P_{in} \qquad (14.25)$$

where P_{in} is the energy of the incident wave and r^* is the complex conjugate of r.

The transmitted portion has an energy

$$P_t = \left|\frac{C}{A}\right|^2 P_{in} \qquad (14.26)$$

The damping produced by the obstacle can thus be written

$$L = \frac{P_{in} - P_t}{P_{in}} = 1 - \left|\frac{C}{A}\right|^2 = 1 - \frac{|C|^2}{(A+B_1)^2}(1+|r|)^2$$

$$= 1 - \frac{|C|^2}{(A+B_1)^2} \frac{4\eta^2}{(1+\eta)^2} \qquad (14.27)$$

The quantities which occur in the expression for L can be measured with a standing-wave detector.

Two examples are:

1. The obstacle is a plane metal short-circuit at right angles to the axis of the waveguide; $E_{tang} = 0$ at this obstacle, and the incident and reflected waves have electric field strengths at the obstacle equal in amplitude but opposite in sign: $B_1 = A$ and $\varphi = \pi$, so that $r = -1$, $\eta \to \infty$ and $C = 0$.

2. The waveguide is open, there are no currents, $H_{trans} = 0$ and E_{trans} is maximum at the obstacle, and the electric field strengths of the incident and reflected waves are equal in magnitude and in phase: $B_1 = A$, $\varphi = 0$, so that $r = +1$, $\eta \to \infty$ and $C = 0$.

14.4 Impedance in waveguides

In the previous sections we have, only by considering the wave-like character of the solutions of Maxwell's equations in waveguides, and with concepts such as the reflection coefficient, standing-wave ratio, short-circuit and open circuit, reached a situation which closely resembles that which we illustrated for parallel-wire transmission lines. The fact that we confined our attention to the principal mode in waveguides does not alter the situation materially; other modes could equally well have been chosen. With the parallel-wire line, the basic variables were current and voltage, while with waveguides we are dealing with electric and magnetic, or electromagnetic, fields. It would thus appear that many phenomena apply equally well to both situations. However, there is one concept encountered with the parallel-wire line that we have not yet come across in connection with the waveguide: the impedance at a given point on the guide.

In the parallel-wire line, the impedance Z was defined as V/I, the ratio of voltage and current, which sufficed to fix Z unambiguously. In a given cross-section, the quantities V and I were independent of position in a lossless parallel-wire line. The current I was the total current through one of the conductors, and

$$V = \int_a^b E_r \, \mathrm{d}r$$

the voltage between the conductors at a point along the line, which was independent of the path of integration as long as the latter was in a transverse plane ($H_z = 0$, the quasi-static case).

Such an unambiguous definition of impedance is impossible for a waveguide: in a transverse cross-section, the current in the wall is a function of position and the line integral of the electric field strength depends on the path along which we integrate in the transverse plane.

However, it would be useful if we could develop a concept of impedance for waveguides. This could be used in various ways, as in Smith charts, for studying impedance transformations along the guide, input impedance, etc. Furthermore, it would be helpful if we could work out an electrical impedance equivalent circuit for representing the effect of obstacles in waveguides at various distances.

There are thus good grounds for developing an impedance concept for use with waveguides analogous to that used for parallel-wire lines, even if our definition cannot be so unambiguous. The fact that quantities such as the reflection coefficient can be used in both cases encourages us to persevere in our attempts.

As we shall see in detail below, doubt will always exist about the absolute value of impedance in a waveguide, but this will not usually be a serious drawback. We may recall that in many cases (Smith charts, etc.) all that is needed is a relative impedance (expressed, for example, in relation to the characteristic impedance).

In order to build up the concept of impedance in waveguides similar to that in parallel-wire transmission lines, we start with the expressions found for the propagation of waves along lossless parallel-wire lines

$$V = A \, \mathrm{e}^{-\mathrm{j}\beta z} + B \, \mathrm{e}^{\mathrm{j}\beta z}$$

$$I = \frac{A}{Z_0} \mathrm{e}^{-\mathrm{j}\beta z} - \frac{B}{Z_0} \mathrm{e}^{\mathrm{j}\beta z} \tag{14.28}$$

The characteristic impedance of the line was $Z_0 = (Z/Y)^{\frac{1}{2}}$, and the propagation constant $\mathrm{j}\beta = (ZY)^{\frac{1}{2}}$, where Z and Y are the impedance and admittance per unit length of line respectively. In general, Z and Y were determined by

$$Z = R + j\omega L$$
$$Y = G + j\omega C$$

from which it follows that

$$Z = j\beta Z_0 = \Gamma Z_0$$
$$Y = \frac{j\beta}{Z_0} = \frac{\Gamma}{Z_0} = \Gamma Y_0 \tag{14.29}$$

V and I were thus the solutions of the telegraphers' equations

$$\frac{\partial V}{\partial z} = -ZI = -\Gamma Z_0 I$$
$$\frac{\partial I}{\partial z} = -YV = -\Gamma Y_0 V \tag{14.30}$$

Now we have already encountered the propagation constant for the TE_{01} mode in a waveguide; it is given by

$$\Gamma = j\beta = \sqrt{\left(\frac{\pi}{a}\right)^2 - \omega^2 \varepsilon \mu} = j\omega\sqrt{\varepsilon\mu}\sqrt{1 - \left(\frac{\lambda}{\lambda_c}\right)^2} \tag{14.31}$$

but we have not yet defined Z or Y, or Z_0.

We must now make a reasonable choice of V and I for the waveguide, in such a way that the analogy with the parallel-wire transmission line is retained and that various important concepts remain valid. It would seem desirable to keep V proportional to the transverse electric field ($V \propto E_t$), so that in a short-circuit plane where $E_t = 0$ we also have $V = 0$. Similarly, the current I should be a linear function of the transverse magnetic field ($I \propto H_t$), so that in an open guide where $H_t = 0$ we also have $I = 0$.

We have thus introduced two proportionality constants, which are still arbitrary in principle. When two waveguides are joined, we can even choose different proportionality constants in the two guides, as long as no discontinuity is produced in the power, VI^*, passing through the junction. We are thus left with one degree of freedom in our definition of impedance, one arbitrary proportionality constant, which we shall now choose so that the analogy with parallel-wire transmission lines is preserved as closely as possible. This is, in fact, the most common approach, although it is of course possible to choose an alternative proportionality constant. If the current and the voltage are chosen, then the impedance and the power are determined, but one may also choose current and power, or voltage and power, as the variables.

The procedure normally used is as follows. The voltage $V(z)$ is chosen proportional to the line integral of the electric field along a straight line

from the middle of the bottom of the waveguide to the middle of the top
(see Fig. 14.5—we assume, as above, that the top and bottom are wider
than the sides)

$$V(z) = q \int_0^b E_x \, dx = q \int_a^b E_0 \sin \frac{\pi}{a} y \, e^{j\omega t - \Gamma z} \, dx = qbE_0 \, e^{j\omega t - \Gamma z} \quad (14.32)$$

The proportionality constant q will be defined later.

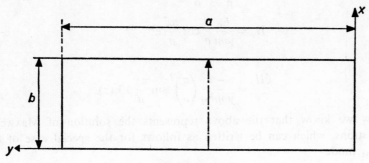

Figure 14.5

For the current, we choose a quantity proportional to the current
density in the z direction in the middle of one of the wider sides of the
waveguide (Fig. 14.6). The current density in the wall is equal to the
discontinuity in the transverse magnetic field at the wall

$$I(z) = p\{H_y(z)\}_{\substack{x=b \\ y=a/2}} = p \frac{\Gamma}{j\omega\mu} E_0 \, e^{j\omega t - \Gamma z} \quad (14.33)$$

The proportionality constant p will again be defined later.

Another choice sometimes made is the total current in the wider wall
of the waveguide:

$$\int_0^a H_y \, dy$$

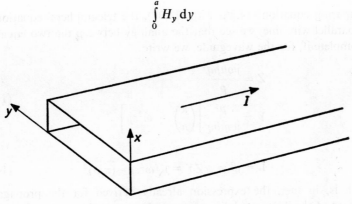

Figure 14.6

We can now express the electromagnetic field of the TE_{01} mode in terms of our new variables $V(z)$ and $I(z)$, as defined in equations (14.32) and (14.33)

$$E_x = \frac{1}{qb} \sin \frac{\pi y}{a} V(z)$$

$$H_y = \frac{1}{p} \sin \frac{\pi y}{a} I(z) \qquad (14.34)$$

$$H_z = \frac{E_0}{j\omega\mu} \frac{\pi}{a} \cos \frac{\pi y}{a} e^{j\omega t - \Gamma z}$$

or

$$\frac{\partial H_z}{\partial y} = \frac{-1}{j\omega\mu bq} \left(\frac{\pi}{a}\right)^2 \sin \frac{\pi y}{a} \cdot V(z)$$

Now we know that the above represents the solution of Maxwell's equations, which can be written as follows for the special case of the TE_{01} mode

$$\frac{\partial E_x}{\partial z} = -\mu \frac{\partial H_y}{\partial t} = -j\omega\mu H_y$$

$$\frac{\partial H_z}{\partial y} - \frac{\partial H_y}{\partial z} = j\omega\varepsilon E_x \qquad (14.35)$$

If we now substitute equation (14.34) in (14.35) in order to express Maxwell's equations in terms of $V(z)$ and $I(z)$, we find

$$-\frac{\partial V(z)}{\partial z} = \frac{j\omega\mu bq}{p} I(z)$$

$$-\frac{\partial I(z)}{\partial z} = \frac{p}{j\omega\mu bq} \left[\left(\frac{\pi}{a}\right)^2 - \omega^2\varepsilon\mu\right] V(z) \qquad (14.36)$$

Comparing equation (14.36) with (14.30), the telegraphers' equation for the parallel-wire line, we see that the analogy between the two equations is complete if, for the waveguide, we write

$$Z = \frac{j\omega\mu bq}{p}$$

$$Y = \frac{p}{j\omega\mu bq} \left[\left(\frac{\pi}{a}\right)^2 - \omega^2\varepsilon\mu\right] \qquad (14.37)$$

or

$$\Gamma = j\beta = \sqrt{ZY} = j\sqrt{\omega^2\varepsilon\mu - (\pi/a)^2} \qquad (14.38)$$

which is, in fact, the expression already derived for the propagation constant of the TE_{01} mode in a rectangular waveguide.

After some manipulation we can also derive

$$Z_0 = \sqrt{\frac{Z}{Y}} = \frac{q}{p} b \sqrt{\frac{\mu}{\varepsilon}} \frac{1}{\sqrt{1-(\lambda/\lambda_c)^2}} \tag{14.39}$$

If the medium in the waveguide is air, we find

$$Z_0 = \frac{q}{p} b \sqrt{\frac{\mu_0}{\varepsilon_0}} \frac{1}{\sqrt{1-(\lambda/\lambda_c)^2}} = \frac{q}{p} b \sqrt{\frac{\mu_0}{\varepsilon_0}} \frac{\lambda_g}{\lambda} \tag{14.40}$$

for the characteristic impedance of a rectangular waveguide for the TE_{01} mode, as derived by analogy with the impedance for parallel-wire transmission lines. The constant q/p still remains to be defined.

We have yet to satisfy the continuous flow requirement of power across a junction between two different waveguides. The power can be expressed as VI^*, and by Poynting's theorem (see Section 5.3)

$$\iint P_z \, dS = \iint E_x H_y^* dx \, dy \tag{14.41}$$

where E, H, V and I are taken as the r.m.s. values.

We can thus put

$$\iint E_x H_y^* \, dx \, dy = VI^*$$

$$\tag{14.42}$$

or

$$E_0^2 \frac{\beta}{\omega\mu} \int_0^b dx \int_0^a \sin^2 \frac{\pi y}{a} \, dy = pqbE_0^2 \frac{\beta}{\omega\mu}$$

or

$$pq = \frac{a}{2} \tag{14.43}$$

We now choose

$$q = 1 \tag{14.44}$$

so that the voltage is given by the line integral of equation (14.32) itself. It follows then that $p = a/2$.

The characteristic impedance of the air-filled waveguide can thus be written

$$Z_0 = \sqrt{\frac{\mu_0}{\varepsilon_0}} \frac{2b}{a} \frac{\lambda_g}{\lambda} \tag{14.45}$$

Most waveguides are dimensioned so that $a \approx 2b$. Since

$$(\mu_0/\varepsilon_0)^{\frac{1}{2}} = 120\pi \approx 377$$

the characteristic impedance becomes $Z_0 \approx 377\lambda_g/\lambda$.

The ratio λ_g/λ normally varies between 1 and 2 in waveguides, so that Z_0 as defined here will be of the order of some hundreds of ohms. When V and I are chosen in other ways, the characteristic impedance is generally of this value, though not too much importance should be attached to this: q may, in principle, be given very extreme values.

We see from the above that the characteristic impedance of the waveguide for the TE_{01} mode is proportional to the height b of the guide. The relationship between Z_0 and the width a is more complicated, because a also occurs in λ_c.

The concept of waveguide impedance can be used in calculations in much the same way as demonstrated previously for parallel-wire transmission lines.

A section of waveguide of length l, terminated by an impedance $Z(l)$ at $z = l$, has an input impedance Z_i which can be expressed as follows in terms of the characteristic impedance Z_0

$$\frac{Z_i}{Z_0} = \left\{\frac{Z(l)}{Z_0} + j \tan \beta l\right\} \Big/ \left\{1 + j \frac{Z(l)}{Z_0} \tan \beta l\right\} \qquad (14.46)$$

while the reflection coefficient caused by $Z(l)$ is

$$r = \left(\frac{Z(l)}{Z_0} - 1\right) \Big/ \left(\frac{Z(l)}{Z_0} + 1\right) \qquad (14.47)$$

For example, the input impedance of a short-circuited section of waveguide of characteristic impedance Z_0 and length l is

$$Z_i = jZ_0 \tan \beta l$$

The quarter-wave impedance transformer, which we have already mentioned in connection with coaxial lines, is also widely used in waveguides (Fig. 14.7). The input impedance $Z_i = Z_{01}$ if $Z_0^2 = Z_{01} Z_{02}$ and $l = \lambda_g/4$. If all three waveguide sections have the same width a, this condition can be satisfied by making $b_0^2 = b_1 b_2$. Of course, if the difference in height between the different sections is too great, parasitic impedances will be produced here too.

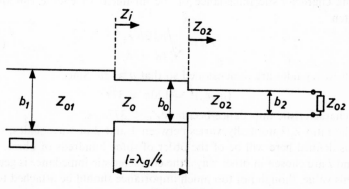

Figure 14.7

The Smith chart can also be used for waveguides, but it should be remembered that the propagation constant β differs from that in free space

$$\beta = \frac{2\pi}{\lambda_g}$$

where $\lambda_g = \lambda/[1-(\lambda/\lambda_c)^2]^{\frac{1}{2}}$.

14.5 Attenuation in waveguides

So far, we have assumed waveguides to be lossless, as we shall in further sections. However, it must be pointed out at this stage that losses do occur in practice. There will be losses in the walls of the guide, and there may also be dielectric losses in the medium filling the guide. The losses found in practice are generally about 25% more than those calculated on the assumption that the waveguide material has a certain finite conductivity (the frequency is generally taken as $f = 1 \cdot 5f_c$ for the purposes of such calculations). The attenuation can be indicated by the real part of the complex propagation constant $\Gamma = \alpha + j\beta$. We then find:

$$\alpha = \frac{R_s \left[1 + \frac{2b}{a} (1 - G^2) \right]}{bG \sqrt{\frac{\mu_0}{\varepsilon_0}}} \quad \text{neper per metre} \qquad (14.48)$$

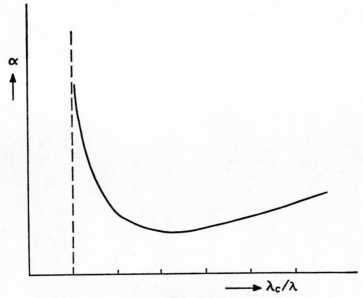

Figure 14.8

where

$$G = \sqrt{1 - \frac{\pi^2}{\omega^2 \varepsilon_0 \mu_0} \left(\frac{m^2}{a^2} + \frac{n^2}{b^2} \right)} = \sqrt{1 - \left(\frac{\lambda}{\lambda_c} \right)^2} \quad (14.49)$$

and R_s is the surface resistance

$$R_s = \sqrt{\frac{\pi f \mu}{\sigma}} = \frac{1}{\sigma \delta} \quad (14.50)$$

where σ is the conductivity ($5 \cdot 8 \times 10^7$ S m^{-1} for copper) and the other symbols have the same significance as in previous sections; δ is the skin depth.

As $\lambda \to \lambda_c$, wall losses increase sharply, as can be seen from equations (14.48) and (14.49). On the other hand, losses increase with decreasing wavelength owing to the skin effect. A general picture of the situation is illustrated in Fig. 14.8. The following loss factors can be used to describe the behaviour of various materials compared with that of copper (1·00): silver 0·97; 70–30 brass 1·92; aluminium 1·25; chromium 1·23; gold 1·19.

15 WAVE PROPAGATION IN CIRCULAR WAVEGUIDES

Waveguides with other than rectangular cross-sections, such as circular ones, are sometimes used. The wave equations can then be solved with their own particular boundary conditions along the same lines we have described above for rectangular guides, by separating the variables and resolving the wave into elementary modes. When dealing with circular waveguides, it is preferable to use cylindrical coordinates r, φ and z as shown in Fig. 15.1. The boundary conditions can then be written

$$
\begin{aligned}
E_\varphi &= 0 \quad \text{at } r = a \\
E_z &= 0 \quad \text{at } r = a
\end{aligned}
\tag{15.1}
$$

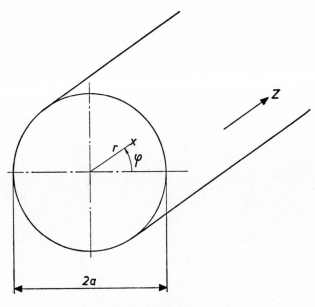

Figure 15.1

One of the differential equations produced in this way turns out to be Bessel's equation, with Bessel functions for its solution instead of the normal trigonometric functions found with rectangular guides. This is not the place to deal fully with circular waveguides; however, the theories involved are basically the same as those developed above for rectangular waveguides.

The principal mode in a circular waveguide, i.e. the lowest mode that can occur independently, is the TE_{11} mode with cut-off wavelength

$$\lambda_c = 2\pi a/1 \cdot 841 \qquad (15.2)$$

Then we have the TM_{01} mode, with

$$\lambda_c = 2\pi a/2 \cdot 405 \qquad (15.3)$$

One mode which is widely used in these circular waveguides is the TE_{01} mode, with

$$\lambda_c = 2\pi a/3 \cdot 832 \qquad (15.4)$$

With this mode, the attenuation actually decreases with increasing frequency; however, a disadvantage is that many other modes (TE_{11}, TM_{01}, TE_{21} and TM_{11}) are likely to occur at the same time.

16 HIGHER MODES IN COAXIAL LINES

It is, of course, possible to solve Maxwell's equations for coaxial lines in the same way as we have done for waveguides. The reason why the telegraphers' equations lead only to the TEM mode is that this equation is derived on the assumption that only transverse electromagnetic fields are involved.

We shall not go fully into the method of solution but will merely mention that more complicated field configurations than the TEM mode are also possible in coaxial lines if the transverse dimensions satisfy certain conditions, similar to those for waveguides. As in the previous chapter, cylindrical coordinates r, φ and z are used, and the solution of the wave

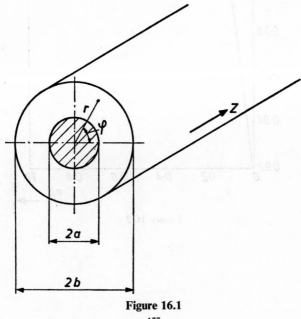

Figure 16.1

177

equation again involves Bessel functions (Fig. 16.1). By way of illustration, the boundary conditions are

$$E_\varphi = 0 \quad \text{at } r = a \text{ and } r = b$$
$$E_z = 0 \quad \text{at } r = a \text{ and } r = b$$

(16.1)

The first mode after the TEM mode is the TE_{10} mode with cut-off wavelength

$$\lambda_c = \frac{(a+b)\pi}{p}$$

(16.2)

where p is approximately equal to 1, so that this mode (and all higher ones) can be avoided by making $(a+b)\pi \lesssim 0.9\lambda$, i.e. by making the average circumference somewhat less than the wavelength. In fact, p is a function of a/b; its form is shown in Fig. 16.2.

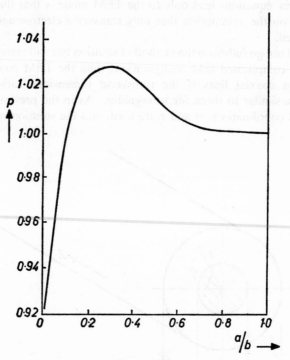

Figure 16.2

17 WAVEGUIDE COMPONENTS

17.1 Discontinuities

The cross-section of a waveguide rarely remains uniform from the source to the load. Many kinds of discontinuities can occur, such as transition from a rectangular to a circular guide, transition between a waveguide and a coaxial line, differences in height, attenuators, diaphragms used to provide a filter effect, and surface roughness left from the finishing process. As we have already seen, the electromagnetic field of the principal mode can no longer satisfy the boundary conditions at such points, and higher modes are produced locally but disappear again at some distance from the discontinuity. The mathematical treatment of such problems is very complicated, and an exact solution is not always possible, except in certain cases where the geometry is extremely simple. Approximate methods, which are often quite complicated, must normally be used to deduce the field configuration near such a discontinuity. Fortunately, we are not usually interested in the short-range effect of discontinuities, but rather in the effect at some distance. The overall effect can be described with the aid of complex reflection and transmission coefficients (see Section 14.3). If the discontinuity has an appreciable length in the axial direction of the waveguide, we must of course specify a reference plane, i.e. the plane at right angles to the axis at which the coefficients describing the effect of the discontinuity have been determined. If necessary, these coefficients can be transformed to refer to other reference planes.

Apart from reflection and transmission coefficients, equivalent impedance networks can also be used to represent the effect of a disturbance, which we imagine to be localized at some distance in a certain reference plane. In equivalent circuits, the obstacle is replaced by a T or π network of impedances in a uniform waveguide, which has the same or nearly the same effect as the obstacle itself when observed from some distance. Such an equivalent circuit is, of course, based on the equivalence of the waveguide and the parallel-wire transmission line, and on the related

179

concept of impedance which we developed for the waveguide in Section 14.4.

A waveguide of characteristic impedance Z_0 and an arbitrary discontinuity (Fig. 17.1) is replaced by one of the same characteristic impedance, together with an impedance network (T or π) situated in a reference plane chosen so that the effect of the obstacle is correctly represented, as shown in Fig. 17.2. The reference plane is often chosen in the middle of the discontinuity, especially if it forms a plane of symmetry of the latter. In this case $z_{11} - z_{12} = z_{22} - z_{12} = z_1$ and $z_{12} = z_2$ or $y_{11} - y_{12} = y_{22} - y_{12} = y_2$ and $y_{12} = y_1$.

For the sake of completeness, Fig. 17.3 sketches a more general form of passive impedance network (a generalized four-pole) in which the currents and voltages are linearly related as follows

$$V_1 = I_1 z_{11} + I_2 z_{12}$$
$$V_2 = I_1 z_{21} + I_2 z_{22} \tag{17.1}$$

or

$$I_1 = V_1 y_{11} + V_2 y_{12}$$
$$I_2 = V_1 y_{21} + V_2 y_{22} \tag{17.2}$$

The sign conventions for the currents and voltages are chosen so that these quantities are positive in the direction of the arrows shown in Fig. 17.3.

How can we now determine the values of the impedances or admittances shown in Fig. 17.2? We have already mentioned that the impedance of a waveguide is determined apart from a multiplicative constant. We shall therefore express the equivalent impedance in terms of the characteristic impedance of the waveguide, and then try to determine the value of this normalized impedance.

In theory, it should be possible to calculate the impedances of the equivalent circuit from the field configuration around the discontinuity. However, even for simple situations such calculations go beyond the scope of this book.

It should also be possible to determine the various parameters of the equivalent circuit by measurement. The three impedances we are interested in, $(z_{11} - z_{12})/z_0$, z_{12}/z_0 and $(z_{22} - z_{12})/z_0$, can, for example, be determined by means of measuring the input impedance with open output, short-circuited output and matched load

$$z_{\text{open}} = z_{11}$$

$$z_{\text{shorted}} = z_{11} - z_{22} + \frac{z_{12}(z_{22} - z_{12})}{z_{22}} = z_{11} - \frac{z_{12}^2}{z_{22}} \tag{17.3}$$

$$z_{\text{matched}} = z_{11} - z_{12} + \frac{z_{12}(z_{22} - z_{12} + 1)}{z_{22} + 1} = z_{11} - \frac{z_{12}^2}{z_{22} + 1}$$

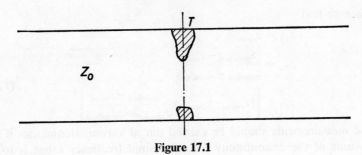

Figure 17.1

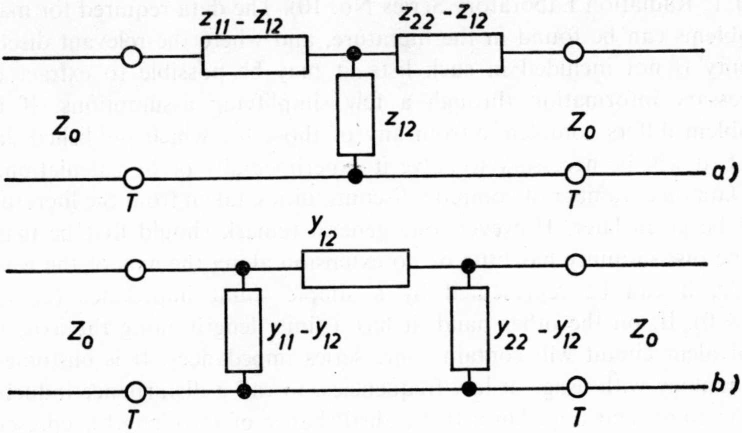

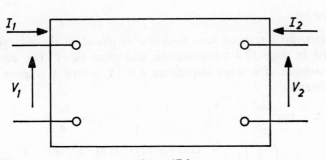

Figure 17.2

Figure 17.3

whence we find

$$z_{11} = z_{open}$$

$$z_{22} = \frac{z_{matched} - z_{open}}{z_{shorted} - z_{matched}}$$

$$z_{12} = \left[(z_{open} - z_{shorted}) \frac{z_{matched} - z_{open}}{z_{shorted} - z_{matched}} \right]^{\frac{1}{2}} \qquad (17.4)$$

These measurements should be carried out at various frequencies if the behaviour of the discontinuity over a desired frequency range is to be determined.

The equivalent circuits for commonly occurring discontinuities have already been determined, either experimentally or by calculation, and are collected in various handbooks (e.g. N. Marcuvitz, *Waveguide Handbook*, M.I.T. Radiation Laboratory Series No. 10). The data required for many problems can be found in the literature, and where the relevant discontinuity is not included in such lists, it may be possible to extract the necessary information through a few simplifying assumptions. If the problem differs completely from any of those for which published data exist, it will be necessary to solve it experimentally or by calculation.

Data on a number of common discontinuities, taken from the literature, will be given later. However, one general remark should first be made. If the discontinuity has little or no extension along the axis of the waveguide, it can be represented by a simple shunt impedance ($z_1 = 0$, $z_2 \neq 0$). If, on the other hand, it has a finite length along the axis, the equivalent circuit will contain some series impedances. It is customary, by analogy with usage at low frequencies, to call a disturbance inductive if the equivalent impedance (for a disturbance of zero length) represents a positive reactance, and capacitive if it represents a negative reactance. We shall assume that the obstacles are made of material of infinite conductivity, so that no resistances will occur in the equivalent circuit. Some examples follow, first of all for obstacles of zero length.

Inductive iris

A symmetrical screen of zero thickness is placed in the waveguide as indicated in Fig. 17.4 (cross-section and plan view). The equivalent circuit consists of a shunt impedance $Z = jX$, where X is given by the approximation

$$\frac{X}{Z_0} \approx \frac{a}{\lambda_g} \tan^2 \frac{\pi d}{2a} \left\{ 1 + \frac{3}{4} \left(\frac{1}{\sqrt{1 - (2a/3\lambda)^2}} - 1 \right) \sin^2 \frac{\pi d}{a} + \right.$$

$$\left. + 2 \left(\frac{a}{\lambda} \right)^2 \left[1 - \frac{4}{\pi} \cdot \frac{E(\alpha) - \beta^2 F(\alpha)}{\alpha^2} \cdot \frac{E(\beta) - \alpha^2 F(\beta)}{\beta^2} - \frac{1}{12} \sin^2 \frac{\pi d}{a} \right] \right\}$$

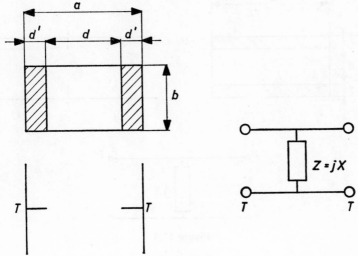

Figure 17.4

where

$$\alpha = \sin\left(\pi d/2a\right) \quad \text{and} \quad \beta = \cos\left(\pi d/2a\right) \qquad (17.5)$$

In the above expression, $F(x)$ and $E(x)$ are complete elliptical integrals of the first and second kind respectively. The expression is valid in a limited wavelength range, and with very thin diaphragms the error is not much more than 1%.

It may often be easier to use the following simpler but somewhat less accurate expressions instead

$$\frac{X}{Z_0} \approx \frac{a}{\lambda_g}\tan^2\frac{\pi d}{2a}\left[1 + \frac{1}{6}\left(\frac{\pi d}{\lambda}\right)^2\right] \qquad \text{for } d/a \ll 1 \qquad (17.6)$$

$$\frac{X}{Z_0} \approx \frac{a}{\lambda_g}\cotan^2\frac{\pi d'}{a}\left[1 + \frac{2}{3}\left(\frac{\pi d'}{\lambda}\right)^2\right] \qquad \text{for } d'/a \ll 1 \qquad (17.7)$$

For $a < 0{\cdot}9\lambda$ and $d < 0{\cdot}5a$ or $d' < 0{\cdot}2a$, the error in X/Z_0 is less than 5% when equations (17.6) and (17.7) respectively are used.

Capacitive screen

Fig. 17.5 gives a cross-section and plan view of screen, together with the equivelent circuit. As above, the reference plane T–T is chosen as the plane of the infinitely thin diaphragm. We find the approximate expressions

$$\frac{X}{Z_0} \approx -\frac{\lambda_g}{4b}\left\{\ln\frac{2b}{\pi d} + \frac{1}{2}\left(\frac{b}{\lambda_g}\right)^2 + \frac{1}{6}\left(\frac{\pi d}{2b}\right)^2\right\}^{-1} \qquad \text{for } d/b \ll 1 \quad (17.8)$$

$$\frac{X}{Z_0} \approx -\frac{\lambda_g}{2b}\left\{\left(\frac{\pi d'}{2b}\right)^2 + \frac{3}{2}\left(\frac{b}{\lambda_g}\right)^2\left(\frac{\pi d'}{2b}\right)^4 + \frac{1}{6}\left(\frac{\pi d'}{2b}\right)^4\right\}^{-1} \qquad \text{for } d'/b \ll 1 \quad (17.9)$$

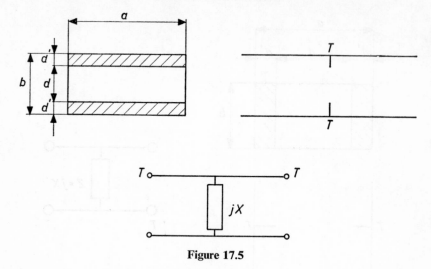

Figure 17.5

When $b/\lambda_g < 0.5$ and $d/b < 0.5$, the error in equation (17.8) is less than 5%, and when $b/\lambda_g < 0.4$ and $d/b > 0.5$, that in equation (17.9) is less than 5%.

Small round hole in otherwise closed screen

The screen is shown in Fig. 17.6. We assume that the hole is made in the middle of the screen. The equivalent shunt reactance is given by

$$\frac{X}{Z_0} \approx \frac{2\pi d^3}{3ab\lambda_g} \quad \text{for } d/b \ll 1 \tag{17.10}$$

The error is less than 10% if $2a > \lambda > 2a/3$, and if the operating wavelength is not too near the cut-off wavelength of the next mode.

Resonant aperture

A 'resonant' screen is placed symmetrically with respect to the middle of a cross-section of the waveguide (Fig. 17.7). In the light of the examples given above, it is not surprising that such an obstacle contains both inductive and capacitive components. This is represented in the equivalent circuit of Fig. 17.7 by means of an LC combination.

The aperture is resonant ($Z/Z_0 \to \infty$) when

$$\frac{a'}{b'} \sqrt{1 - \left(\frac{\lambda}{2a'}\right)^2} \approx \frac{a}{b} \sqrt{1 - \left(\frac{\lambda}{2a}\right)^2} \tag{17.11}$$

We shall now give a few examples of an obstacle of finite extension along the axis of the waveguide, thus giving series components in the equivalent circuit.

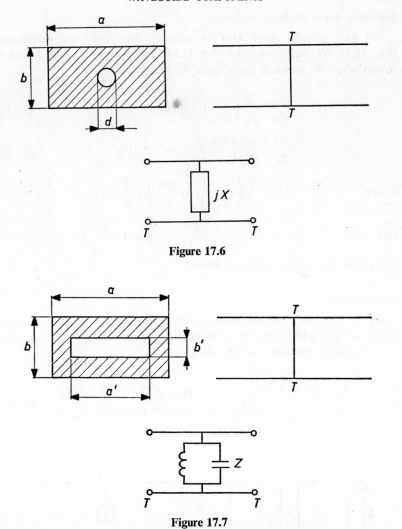

Figure 17.6

Figure 17.7

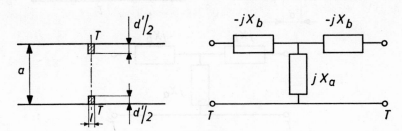

Figure 17.8

Inductive screen of finite thickness

We have already dealt with an infinitely thin inductive diaphragm (Fig. 17.4). We shall now show (Fig. 17.8) what happens to the equivalent circuit when the obstacle has a finite thickness l. We find

$$\frac{X_a}{Z_0} = \frac{2a}{\lambda_g}\left(\frac{a}{\pi D_1}\right)^2, \quad \text{for } \pi D_1/\lambda \ll 1 \tag{17.12}$$

$$\frac{X_b}{Z_0} = \frac{a}{8\lambda_g}\left(\frac{\pi D_2}{a}\right)^4, \quad \text{for } \pi D_2/\lambda \ll 1 \tag{17.13}$$

where
$$D_1 \approx \frac{d'}{\sqrt{2}}\left(1 + \frac{l}{\pi d'} \ln \frac{4\pi d'}{el}\right), \frac{l}{d'} \ll 1$$

and
$$D_2 = \sqrt[4]{\frac{4ld'^3}{3\pi}}, \quad \frac{l}{d'} \ll 1$$

and e the base of natural logarithms.

Inductive rod

Fig. 17.9 shows a rod of circular cross-section, fixed between the mid-points of the wide sides of a rectangular waveguide. The equivalent circuit contains the series components

$$\frac{X_b}{Z_0} \approx \frac{a}{\lambda_g} \frac{\left(\dfrac{\pi d}{a}\right)^2}{1 + \dfrac{11}{24}\left(\dfrac{\pi d}{a}\right)^2} \qquad \begin{array}{l}\text{(For } d/a < 0\cdot 2\text{, the error}\\ \text{is a few per cent.)}\end{array} \tag{17.14}$$

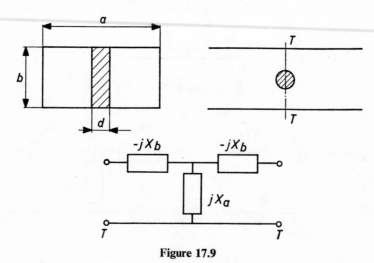

Figure 17.9

$$\frac{X_a}{Z_0} \approx \frac{1}{2}\frac{X_b}{Z_0} + \frac{a}{2\lambda_g}\ln\frac{4a}{\pi d e^2} \qquad (17.15)$$

As d increases, X_b becomes relatively more important, while it is negligible when the diaphragm is thin.

Capacitive rod

Fig. 17.10 shows a rod of circular cross-section clamped between the narrow sides of the waveguide. We give the equivalent circuit as a π network of admittances.

$$\frac{B_a}{Y_0} \approx \frac{2b}{\lambda_g}\left(\frac{\pi D}{2b}\right)^2, \quad \frac{D}{b} \ll 1 \qquad (17.16)$$

$$\frac{B_b}{Y_0} \approx \frac{\lambda_g}{2b}\left(\frac{2b}{\pi D}\right)^2, \quad \frac{D}{b} \ll 1 \qquad (17.17)$$

If $D/b < 0.3$ and $2b/\lambda_g < 0.4$, the error is less than 10%. In these expressions, $Y_0 = 1/Z_0$ is the characteristic admittance of the waveguide.

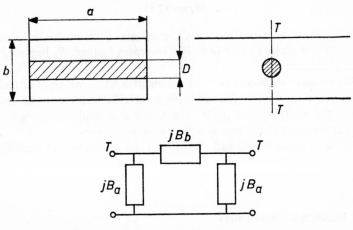

Figure 17.10

Rod of variable length

Our last example is a rod of diameter d and height h, as shown in Fig. 17.11. If $h = b$, the situation is the same as with the inductive rod above.

No generally valid expression for the equivalent impedance as a function of h and d is known. In Marcuvitz's *Waveguide Handbook*, experimental data are given for this situation in 30 mm waveguides. If h is small compared with the height b of the waveguide, $X_a < 0$, i.e. the over-all effect of the obstacle as observed from a distance is capacitive. As h increases, the absolute value of X_a decreases (the system becomes

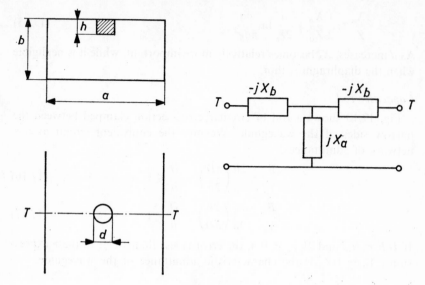

Figure 17.11

less capacitive or more inductive), until series resonance is produced ($X_a = 0$) at a certain value of h. If h increases further, X_a becomes positive, i.e. inductive.

This system can be achieved by means of a screw which is turned into or out of the waveguide, thus allowing a variable reactance to be 'shunted across' the guide at that point. The larger d is made, the larger is X_b, and the narrower is the region within which X_a varies.

All the systems mentioned above, and many others, are used for the construction of filters, impedance transformers, matched loads, etc., in waveguides.

17.2 Impedance transformers

We have already dealt with impedance transformers in our discussion of matching problems and the like in parallel-wire transmission lines. These components are often needed in waveguides too, and we shall discuss a few different types in this section. Now we know already that a uniform section of waveguide (or parallel-wire transmission line) of length l transforms an impedance Z placed at one end to another value, which depends on l (see, for example, equation 14.46). We often make use of this fact when using Smith charts. However, it is usually necessary to have an impedance transformation at a definite point, often where the disturbance is situated or in a plane of symmetry, to be able to match over a wide frequency range.

Tuning screw

This form of impedance transformer (or rather impedance added at a given point in order, for example, to compensate another impedance) has been discussed above (see also Fig. 17.12). Depending on the length h of the tuning screw, we can introduce either a self-inductance or a capacitance in this way. It will be clear that current will flow in the screw

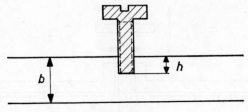

Figure 17.12

and care should therefore be taken to ensure that the screw-thread is of high quality so that the losses are not too high. It is even better to use a choke construction (Fig. 17.13), so that no currents flow in the wall of the waveguide at the site of the short-circuit. The use of more than one tuning screw (for example, two, a distance $\lambda_g/8$ apart along the waveguide) has much the same effect as more than one stub tuner in a coaxial line (see Section 11.5).

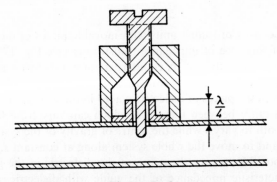

Figure 17.13

Sliding screw

Instead of using several screws a fixed distance apart, we could use one screw which can slide along the waveguide. This can be done in various ways, which we shall not go into here. The depth to which the screw penetrates into the waveguide (the slit, through one of the wide sides of the waveguide, should be kept narrow to avoid stray radiation and its edges should be bevelled to avoid internal reflections) should remain

7*

practically constant as it slides along the waveguide. As above, it is better to use a $\lambda/4$ choke construction for the screw. This system makes it possible to place a reactance corresponding to the screw diameter d and the depth of penetration h (see Section 17.1) at any desired point within the range of the slide. This range generally covers several guide wavelengths (see Fig. 17.14).

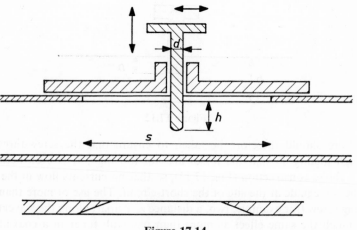

Figure 17.14

Slug tuners

Slug tuners consist of tuning units with movable pieces of dielectric. The operation of this type of impedance transformer (see Fig. 17.15) is completely analogous to that of its counterpart in parallel-wire transmission lines (see Section 11.5), but its construction is somewhat different.

The width of the pieces of dielectric placed in the waveguide is $p = \lambda_g/4$, while the distance between them is s, which can vary from 0 to $\lambda_g/4$. It is possible both to vary s while the centre of gravity of the system remains stationary, and to move the whole system along at constant s. It is sometimes also possible to vary the depth of the dielectric in the guide. If Z_1 is the characteristic impedance of the guide with dielectric, and Z_0 the characteristic impedance without dielectric, the input impedance on the left of the tuner can be made $(Z_1/Z_0)^4$ if the guide is terminated with a matched load to the right of the slug tuner.

Stub tuners

Stub tuners take the form of tuning units in side-arms of the waveguide. The E plane and H plane T junctions shown in Fig. 17.16a and b are not very widely used. These components are so named because the side-arms

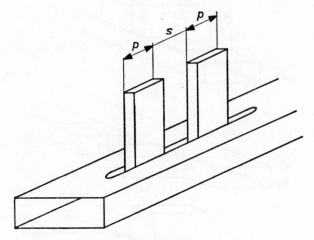

Figure 17.15

are parallel to the *E* field and the *H* field in the main guide respectively. In the first case, the stub impedances are in series with the main guide, and in the second case they are in parallel. The side-arms are provided with short-circuit plungers (not shown in the figure), so that all reactances between $-\infty$ and $+\infty$ can be connected in series or in parallel with the main guide by shifting the plungers at least one half wavelength. The distance between the side-arms again is about $\lambda_g/8$ or $3\lambda_g/8$, to make the tuning range as wide as possible (see also Section 11.5).

The *E–H* tuner shown in Fig. 17.16*c* is more often used. This allows a reactance in series and one in parallel to be introduced at the same point in the waveguide. In order to limit the length of these tuners, types *a* and *b* are often made with the side-arms alternately on opposite sides of the main waveguide. We shall be returning to *T* junctions below.

Quarter-wavelength impedance transformers

These components have already been discussed in Section 14.4 (see Fig. 14.6). An impedance transformer of this type can only match two real impedances. If the output impedance (Z_{02} in Fig. 14.6) is complex, this piece of line should be lengthened to give a real impedance Z_{02} at the transformer.

The required length of the line segment can be determined from the Smith diagram, or from the formula for the input impedance of an arbitrary terminated length of waveguide. It would be convenient if the transformer could be made movable along the guide. While this is possible, it has certain disadvantages in waveguides.

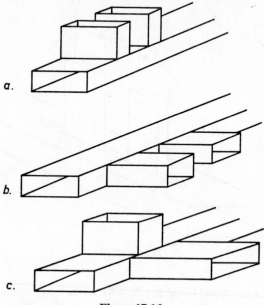

Figure 17.16

In Section 14.4 we did not consider the effect of parasitic impedances (due to discontinuities in the height of the waveguide). In fact, especially when the difference between Z_{01} and Z_{02} is large, parasitic capacitance will not be negligible, so that $Z_0^2 = Z_{01}Z_{02}$ and $l = \lambda_g/4$ will not give matching. This can be corrected by keeping $Z_0^2 = Z_{01}Z_{02}$ and reducing the length l somewhat to compensate for the extra capacitance. In order to do this, of course, we must know the value of this capacitance. For this purpose, we make use of the equivalent circuit of Fig. 17.17 for a waveguide with a discontinuity in the height.

If we call $b'/b = \alpha = 1-\delta$, then we find from waveguide handbooks that the parasitic capacitance due to the discontinuity is approximately given by

$$\frac{B}{Y_0} \approx \frac{4b}{\lambda_g} \left[\ln \frac{e}{4\alpha} + \frac{\alpha^3}{3} + \frac{1}{2}\left(\frac{2b}{\lambda_g}\right)^2 (1-\alpha^2)^4 \right] \quad \text{if } \alpha \ll 1 \qquad (17.18)$$

or

$$\frac{B}{Y_0} \approx \frac{4b}{\lambda_g}\left(\frac{\delta}{2}\right)^2 \left[\frac{2\ln 2/\delta}{1-\delta} + 1 + \frac{17}{16}\left(\frac{2b}{\lambda_g}\right)^2 \right] \quad \text{if } \delta \ll 1 \qquad (17.19)$$

The first expression is accurate to within 5% if $\alpha < 0.6$ and $2b/\lambda_g < 0.5$, and within 2% if $\alpha < 0.4$ and $2b/\lambda_g < 0.4$. The second is accurate to within 5% if $\delta < 0.5$ and $2b/\lambda_g < 0.5$, and within 3% if $\delta < 0.4$ and $2b/\lambda_g < 0.4$.

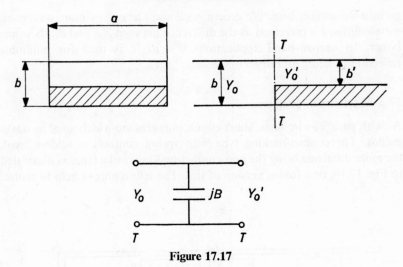

Figure 17.17

Once the parasitic capacitance has been calculated in this way, the line can be shortened to the required extent to allow for this capacitance at each end.

If the two impedances to be matched differ widely, the quarter-wave transformer can be made in steps to avoid excessive discontinuities in height. The bandwidth within which the transformer works properly decreases as the impedance discontinuity increases.

Gradual transition (taper)

We have already come across the taper (Fig. 17.18) as a means of providing a reflection-free connection between guides of different dimen-

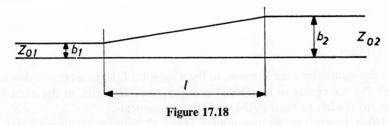

Figure 17.18

sions (width and/or height) in connection with the coaxial line. In wave-guides, as in coaxial lines, the simplest form of taper is that in which the dimensions of the transformer vary linearly. The theoretical determination of the best form (shortest length l for a given maximum permissible reflection coefficient) is even more complicated than for the parallel-wire transmission line (λ_g varies with dimension a of the guide), and we shall not go into the subject here. We generally choose l as one or more waveguide wavelengths; l is increased as the difference between Z_{01} and Z_{02} becomes larger. In narrow-band applications, $l = n\lambda_g/2$ is used for minimum reflection (n being an integer).

17.3 Short-circuit plungers

As with parallel-wire lines, short-circuit plungers are widely used in wave-guides. The contact-making type with spring contacts is seldom used, the more usual one being the non-contact-making choke type, as illustrated in Fig. 17.19, or a folded version of this. The teflon plugs ε help to centre,

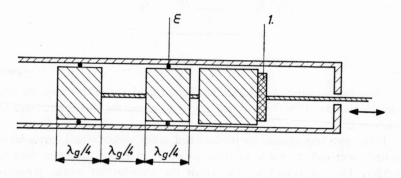

1. damping material

Figure 17.19

the plunger, and make it easier to move. The theory of this plunger (as lengths of rectangular coaxial line) is not fully explained, but it works very well in practice.

17.4 Bends

We distinguish between E bends, in the wide side of the guide (up or down, with the waveguide in its normal position), and H bends, in the narrow side (to the left or right in the normal arrangement).

These bends can be continuous (Fig. 17.20a) or in discrete steps (Fig. 17.20b). Reflection in regular bends can be reduced if the radius of curvature R is large enough. If the mean length of the waveguide round the 90° bend ($\approx \pi R/4$) is a few guide wavelengths, reflections will be fairly low.

Various different kinds of sharp bends can be distinguished. In the first place we have the single sharp bend (Fig. 17.21). If θ is not very small, the parasitic impedances X and B can be quite large, so that appreciable reflection is produced. In certain ranges of θ, resonance occurs ($X \to \infty$ or $B \to 0$). Bends can also be constructed with two sharp discontinuities (Fig. 17.22), when the length L can be chosen so that the reflections from

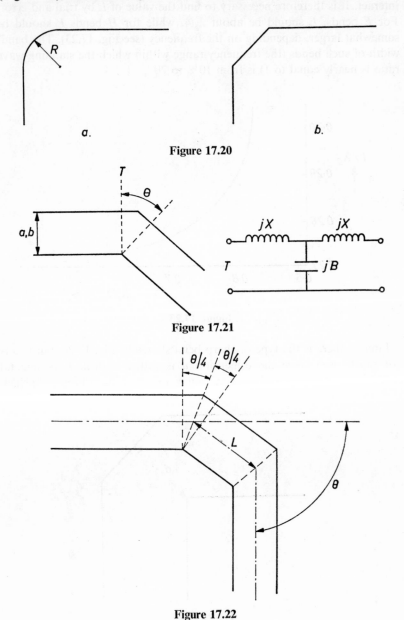

Figure 17.20

Figure 17.21

Figure 17.22

the two discontinuities cancel out. The value of L cannot be determined from the data for the individual discontinuities, because the higher modes produced at the first discontinuity will not have disappeared completely by the time the second is reached; the two discontinuities thus interact. It is therefore necessary to find the value of L by trial and error. For E bends, L should be about $\lambda_g/4$, while for H bends L should be somewhat larger, depending on the frequency (see Fig. 17.23). The bandwidth of such bends (the frequency range within which the standing-wave ratio is nearly equal to 1) is from 10% to 20%.

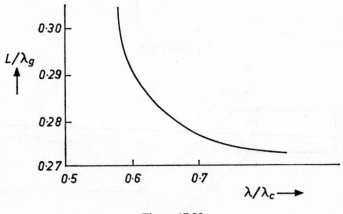

Figure 17.23

Finally, there is the type of sharp bend shown in Fig. 17.24, but this is not much used since the dimension d is rather critical. Experimental curves giving the value of d for minimum reflection have been published.

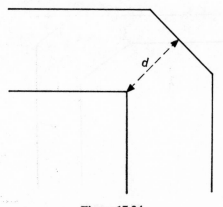

Figure 17.24

17.5 Transitions from waveguide to coaxial line

Transitions from a waveguide to a coaxial line often have to be provided, for example, because generators or detectors have coaxial output plugs, or because certain components such as rotary joints or parts of attenuators are often made in coaxial form.

We shall restrict ourselves here to a TEM mode occurring in the coaxial ine and a TE_{01} mode in the waveguide. The overall field configuration is shown in Fig. 17.25. The coupling can be carried out in two

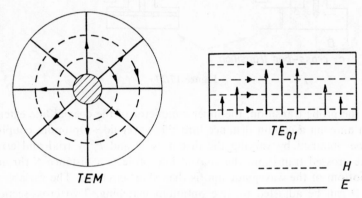

Figure 17.25

ways: electrically, with a probe or antenna for the E lines (Fig. 17.26a); or magnetically, with a coupling loop for the H lines (Fig. 17.26b). Electric coupling is more often used, and various types are available of which a few will be indicated here (see Fig. 17.27).

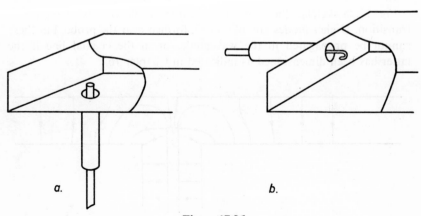

Figure 17.26

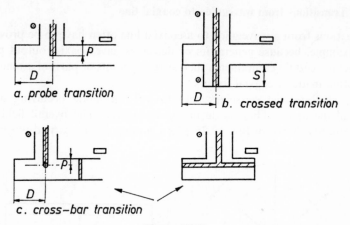

a. probe transition

b. crossed transition

c. cross-bar transition

Figure 17.27

In the probe transition, the inner conductor of the coaxial line extends as an antenna a certain distance into the waveguide. Optimum coupling can be obtained by varying the distances D and P by trial and error. In the crossed transition, the coaxial line projects a distance S through the bottom of the waveguide and is then short-circuited. The variables S and D can be adjusted to give optimum matching. Two cross-sections through a third type, the cross-bar transition, are illustrated. Here optimum coupling is obtained by varying P and D.

Some kind of impedance transformation (diaphragm or inductive or capacitive rod) is often used to reduce the reflection of the transition in a certain frequency band. The dimensions of such transformers are nearly always determined by trial and error (in order to give the lowest possible reflection in the maximum possible frequency range).

Fig. 17.28 sketches the form of the E lines in the vicinity of the probe transition. Higher modes are, of course, formed near the probe, but these cannot be propagated in the waveguide, or in the coaxial line if the latter has been dimensioned as indicated in Chapter 16.

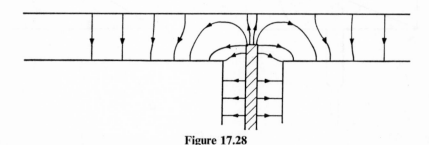

Figure 17.28

17.6 **Detection**

We have already dealt with the detectors normally used for microwave frequencies (thermistors, diodes) in our discussion of coaxial components. Similar detectors are also used in waveguides, though the diode holder, and sometimes the crystal housing, will be designed differently. Sometimes coaxial detectors are used, together with one of the coaxial-waveguide transitions discussed in the previous section, but it is also possible to place the detector (thermistor or diode) directly in the waveguide. The most sensitive point for electric detection is half-way between the middles of the wide sides of the waveguide (Fig. 17.29), where the electric field of the TE_{01} mode is at a maximum. With the aid of a short-circuit plunger, a voltage maximum of the standing-wave pattern is made to fall at the site of

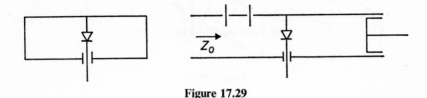

Figure 17.29

the detector, and the detector impedance is matched to the characteristic impedance Z_0 of the waveguide at the operating frequency by means of a few tuning screws or other impedance transformers. Due to this tuning, these types are quite narrow-band in principle. Less critical types can be obtained by sacrificing sensitivity. Since they have less reflection, even without tuning, they can be used over a wider frequency range without adjustment. One possible form is shown in Fig. 17.30. Here the diode is placed nearer the side of the guide. The wave in the guide is not influenced by the impedance of the diode as much as in the previous case. Slight reflections which may be present are easier to match over a wide bandwidth.

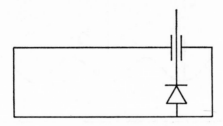

Figure 17.30

The wall-current detector can operate over a very wide band, but is quite insensitive. Basically, here the coaxial diode is placed in a hole in the side of the waveguide (Fig. 17.31).

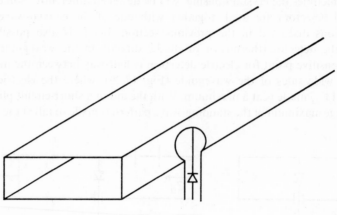

Figure 17.31

17.7 Standing-wave detectors

We have already discussed the standing wave concept and the instrument used to detect such a field pattern. The field is scanned by means of an antenna which can slide along the waveguide (or coaxial line), a detector and a measuring instrument. Standing-wave ratios, impedances and wavelengths (Fig. 17.32) can be measured in this way (see also Section 11.4). It will be obvious that the probe may not influence the field appreciably, resulting in a low penetration depth.

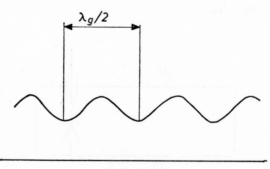

Figure 17.32

In precision instruments, high demands are made on the construction of the slide. When it moves, the probe should not be subject to any vertical or transverse motion, since any displacements would lead to a change in the measured field (voltage) which could not be distinguished from changes due to the standing-wave pattern. It would then be impossible to measure a standing-wave ratio near unity with any accuracy. In order to optimize sensitivity at the penetration depth of the probe in the waveguide, the antenna must be matched to the coaxial line in which the detector is placed. This can be done by placing a number of stub tuners in the coaxial line, as shown in Fig. 17.33.

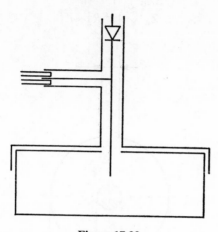

Figure 17.33

Sometimes the antenna is not coupled to a coaxial line, but to an auxiliary waveguide in which the detector is mounted (see Fig. 17.34). The output power is generally so low that the detector reacts quadratically, i.e. the output voltage is proportional to the square of the field strength.

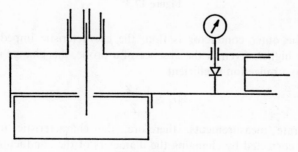

Figure 17.34

Any system involving a sliding probe will, of course, require a slit to be made in the waveguide. This slit must be parallel to the lines of current flow in the surface so that the original field is disturbed as little as possible. In a coaxial line the slit should therefore be parallel to the axis, and in a waveguide it should be in the middle of one of the wide sides and parallel to the axis of the guide (for the TE_{01} mode). The length of the slit is generally a multiple of waveguide wavelengths.

A slit of this kind in a coaxial line (Fig. 17.35) alters the characteristic impedance. If the outer conductor is very thick, the change in the characteristic impedance is given by

$$\frac{\Delta Z_0}{Z_0} \approx \frac{1}{4\pi^2} \frac{w^2}{R_2^2 - R_1^2} \tag{17.20}$$

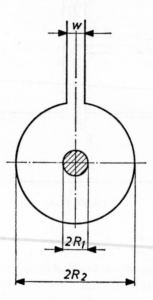

Figure 17.35

while if the outer conductor is thin, the characteristic impedance will be slightly higher. Even in the absence of a probe, this change in Z_0 will give rise to a reflection coefficient

$$|r| \approx \frac{1}{2} \frac{\Delta Z_0}{Z_0} \tag{17.21}$$

For accurate measurements, therefore, the characteristic impedance should be corrected by changing the diameters of the conductors slightly at the site of the slit.

A slit in a waveguide (Fig. 17.36) also causes a change in the characteristic impedance

$$Z_0' \approx Z_0 \left(1 + \frac{w^2 \lambda_g^2}{8\pi a^3 b} \right) \qquad (17.22)$$

and also in the waveguide wavelength

$$\lambda_g' \approx \lambda_g \left(1 + \frac{w^2 \lambda_g^2}{8\pi a^3 b} \right) \qquad (17.23)$$

We can also speak of an apparent narrowing of the guide

$$a' = a - \frac{w^2}{2\pi b} \qquad (17.24)$$

The above expressions still hold approximately for a finite thickness of the walls of the waveguide, as long as this thickness is of the same order as the width of the slit. The change in the waveguide wavelength (which can be measured directly with the standing-wave detector) is generally less than 1%.

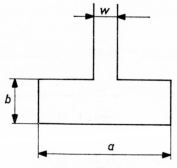

Figure 17.36

If the wall thickness is greater than or equal to the slit width, the radiation through the slit is also slight.

When the system is asymmetrical, the presence of the probe can lead to higher modes, which can sometimes be propagated in the slit and lead to errors in the measurement. These modes are not generated if the system is completely symmetrical. The effect of the probe on the field pattern does not entail any significant change in the minima, although its reactive components do shift the maxima slightly. To determine the correct phase of the standing wave, therefore, one should preferably measure the position of the minimum (by means of symmetrical measurements on both sides of the minimum).

17.8 Phase shifters

A phase shifter is an instrument for varying the (electric) length of the line by mechanical means. It is not generally possible to vary the actual length of a waveguide by means of sliding waveguide segments.

The phase shifter is generally used in combination with a fixed impedance transformer to resolve matching problems. There are various types: Fig. 17.37 shows one form, in which long slits are made in the wide sides of the waveguide. If the narrow sides of the waveguide are squeezed together by a suitable mechanism, the waveguide wavelength will be changed locally

$$\lambda_g = \lambda/\sqrt{1-(\lambda/2a)^2}$$

This involves a change in the phase velocity, and hence in the effective (electric) length of the section of waveguide in question. The slit is made so long that when the guide is squeezed to its maximum extent the dimensions only change very gradually in order to reduce reflection.

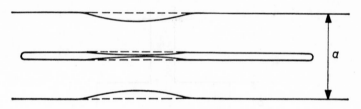

Figure 17.37

Another form of phase shifter is shown in Fig. 17.38. A sheet of dielectric (tapered to prevent reflections) can be moved from the side of the waveguide to the middle. The supports should cause as little reflection as possible by keeping them thin and by having, for example, two supports an average distance of $3\lambda_g/4$ apart. As the dielectric is moved towards the middle of the guide, where the electric field strength is higher, the effect of the dielectric increases. For a given required phase shift (where the phase velocity in the dielectric is different from that in the empty guide) we determine experimentally both the total length and thickness of the dielectric (of given dielectric constant) and the length of the tapering.

Besides the above type, in which the dielectric is always inside the waveguide, we can also make a modified form in which the dielectric is introduced a variable distance into the waveguide through a slit in the middle of a wide side.

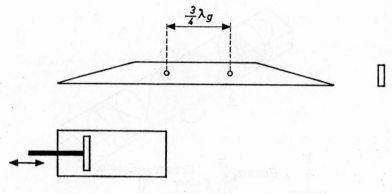

Figure 17.38

17.9 Attenuators and matched loads

As with coaxial lines, waveguides have both resistance attenuators and cut-off attenuators. Coaxial cut-off attenuators can be used in waveguides by connecting them at both ends by waveguide-coaxial transitions. However, resistance attenuators are more often used, except where the attenuator is used as a filter (we shall return to this later).

The resistance attenuator can be made along the same lines as the phase shifter shown in Fig. 17.38, or by inserting a piece of material from the outside through a slit in the middle of a wide side. For an attenuator, of course, the material used is chosen for its attenuating properties (e.g. graphitized Pertinax). A fixed attenuator can be made by mounting a piece of lossy material in a fixed position in the waveguide. The attenuating discs are again tapered in the longitudinal direction in order to avoid reflections. An attenuation of at least 20 to 30 dB can be achieved in this way with quite low reflection coefficients.

There is another type of attenuator which can be very accurately adjusted; namely, the rotary attenuator (see Fig. 17.39, reproduced from *Philips Technical Review*, **22**, 4, p. 129, 1960). A section of waveguide of circular cross-section is placed between two lengths of rectangular waveguide through two gradual transitions. A thin sheet of resistive material is mounted in this round section in such a way that it can rotate about an axis which coincides with the axis of the round waveguide. A fixed resistive sheet is mounted on each side of the rotating sheet, the two fixed sheets being coplanar. The waveguide is designed so that the electric field E of the wave is normal to the first fixed damping sheet. Now of course components perpendicular to the damping sheet are not attenuated (the field does not produce any current). Let us suppose that the rotating

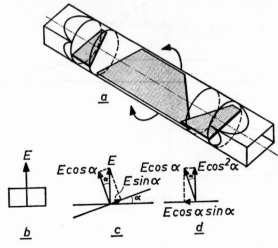

Figure 17.39

sheet makes an angle α with the fixed sheets. We resolve the electric field into a component perpendicular to the rotating sheet and one in its plane. If the rotating sheet is long enough, the component parallel to it will be completely attenuated and the amplitude of the field after this middle sheet will be $E \cos \alpha$. We can again resolve this field into a component perpendicular to the last, fixed damping sheet and one parallel to it. At the end of the last fixed sheet (again, if it is long enough), the amplitude of the remaining field is $E \cos^2 \alpha$. The total power attenuation is thus proportional to $\cos^4 \alpha$. If only relative damping is involved, the attenuation produced can thus be read off directly without calibration. For absolute measurements, the zero damping of the attenuator as a whole should be known.

A matched load can be provided by a piece of damping material (wood, compressed fibre, graphite, etc.) of gradually increasing transverse dimensions, as described above for the coaxial line.

17.10 Directional couplers

For the applications and principle of operation of direction-dependent couplers, the reader is referred to Section 12.6, where these components are discussed in their coaxial form. It may be added here that a directional coupler can also be used as a fixed attenuator with a maximum attenuation of a few tens of decibels. For higher attenuations, the demands made on the matched loads in the auxiliary line are too high. We shall now discuss a number of quantities characteristic of directional couplers, with reference to Fig. 17.40.

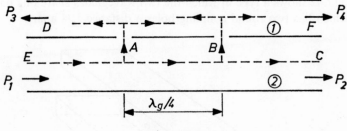

Figure 17.40

The directional coupler basically consists of a main line (2) and an auxiliary or side line (1), the two being coupled by means of coupling holes A and B an electric distance $\lambda_g/4$ apart. The optimum distance between the coupling holes may be determined by trial and error.

The amount of energy coupled into the side line is expressed by the *coupling factor* $10 \log P_1/P_4$ dB, if matched loads are provided at C and D. The attenuation of the wave incident at E, in direction D of the side line, is given by the *directivity* $10 \log P_4/P_3$ dB, if the main line is provided with a matched load at C.

The coupling factor naturally depends on the strength of the coupling, and can have a value of between 3 and 20 or 30 dB in normal cases. The directivity often has a value of 30–35 dB, which is adequate for most applications.

Apart from differences in coupling factor, the side line can be designed in a number of different ways. Often only one of the outputs D or F is accessible, the other being terminated with a matched load. In a dual design both outputs are accessible for measurements of the standing-wave ratio. We shall now illustrate a number of types that differ in the way in which the coupling is effected.

Fig. 17.41 shows one type in which the two waveguides are coupled by two quarter-wavelength guide sections of height b' and normal width. The coupling factor is approximately $20 \log b/b'$.

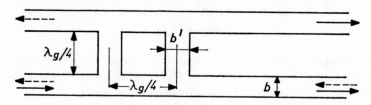

Figure 17.41

Fig. 17.42 shows a type in which the two waveguides share one narrow side, which has two round coupling holes a distance $\lambda_g/4$ apart.

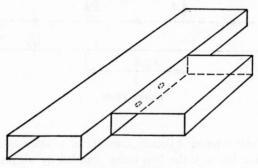

Figure 17.42

Fig. 17.43 shows how the coupling factor can be increased in this type with two identical coupling devices, by replacing the small round coupling holes by slits of length about $\lambda_g/4$ in the common wall. The width d and the distance s for optimum coupling should be determined by trial and error.

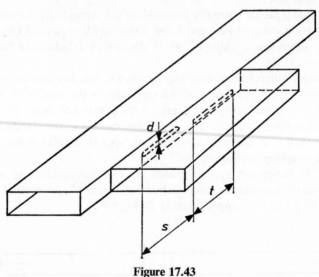

Figure 17.43

Directional couplers are also made with a larger number of identical coupling holes, instead of only two. The coupling can be made stronger in this way, but the system is more frequency-dependent.

A directional coupler can also be made with one long, narrow slit in the common wall. If the slit is made so that the width w gradually decreases towards the ends, thus reducing reflections, the directivity can be made quite high even though the slit is quite wide (large w) and hence the coupling factor low. The slit has an optimum length l above which the coupling factor increases again due to feedback from the side line to the main line. For low coupling (high coupling factor), the power coupled into the side line is proportional to l^2 and w^6. The optimum dimensions are found by trial and error.

17.11 T-junctions

In this section we shall discuss a number of waveguide T-junctions commonly used for a variety of applications (for example, in bridge circuits or as power dividers).

Firstly, two types of T-junctions with three ports: Fig. 17.44 shows an E plane T, where the axis of the side arm is parallel to the E plane in the main waveguide; and Fig. 17.45 illustrates the form of the electric field in the E plane T. This figure does not show the over-all field pattern at any moment, but the track of a line of force when the T is fed from the side arm. If the T is made symmetrical, the energy will be divided equally between the right-hand and left-hand branches of the main guide, but the electric field in these two branches will be in anti-phase.

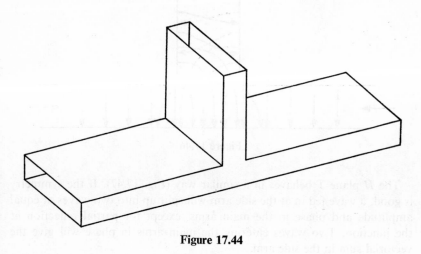

Figure 17.44

Fig. 17.46 shows what happens when the T is fed by two waves of equal amplitude and phase in the two main branches. If the T is completely symmetrical, the two waves in the side arm will be equal and in anti-phase, and will thus cancel out. The resultant wave will consequently be zero. In general, with an arbitrary relationship between the amplitude and phase of the two waves, the field in the side arm will represent the vectorial difference between the fields coming from the two main branches.

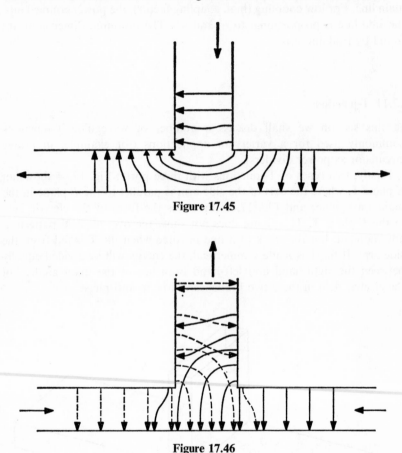

Figure 17.45

Figure 17.46

The *H* plane T behaves in a similar way (Fig. 17.47). If the symmetry is good, a wave fed in at the side arm will split up into two waves of equal amplitude and phase in the main arms, except for partial reflection at the junction. Two waves entering the main arms in phase will give the vectorial sum in the side arm.

An equivalent impedance circuit can be constructed for such T-junctions. Even if the T is highly symmetrical, the reflection coefficient in one arm will not be zero when the two other arms are terminated with a matched load. Fig. 17.48 shows the equivalent circuit of an *E* plane T.

The literature gives the following approximate values for the impedance parameters

$$\frac{B_a}{Y_0} \approx \frac{\pi b}{\lambda_g}\left(0\cdot841+2\cdot04\,\frac{b^2}{\lambda_g^2}\right) \tag{17.25}$$

$$\frac{B_b}{Y_0} \approx 0\cdot859\,\frac{b}{\lambda_g} \tag{17.26}$$

$$\frac{B_c}{Y_0} \approx \frac{\lambda_g}{2\pi b} \tag{17.27}$$

$$\frac{B_d}{Y_0} \approx 1\cdot56\,\frac{b}{\lambda_g} \tag{17.28}$$

This E plane T is also called a series T, because of its equivalent circuit, while the H plane T, whose equivalent circuit is shown in Fig. 17.49, is called a shunt T. The approximate expressions for X_a, X_b, X_c and X_d for the shunt T are much more complicated, because of the dependence on $\lambda_g/2a$. We shall not give these formulae here.

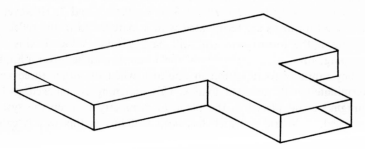

Figure 17.47

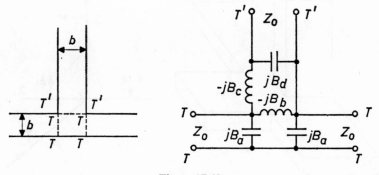

Figure 17.48

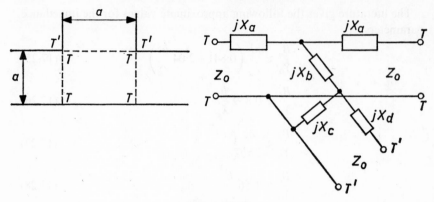

Figure 17.49

The hybrid T, or *E–H* T-junction, shown in Fig. 17.50, which is a combination of the *E* and *H* T-junctions, is widely used. In the light of what we have seen above about the properties of the *E* and *H* T-junctions separately, it will be clear that when the symmetry of the hybrid T is sufficiently close, the series and parallel arms will be decoupled, i.e. there is no direct coupling between *P* and *S* or between 1 and 2. However, if three of the four arms are terminated with a matched load, the reflection coefficient in the fourth arm will still have a finite value. If this is a disadvantage for certain applications, the T can be matched over a limited frequency range (zero reflection in one arm when the other three arms are terminated with matched loads). This matching is achieved with a diaphragm or rod placed in the plane of symmetry to keep the system symmetrical. A hybrid T which has been matched in this way is called a 'magic T'.

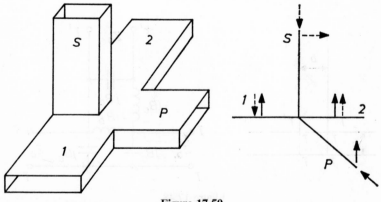

Figure 17.50

One important application of the hybrid T is in bridge circuits for measuring unknown impedances by comparison with calibrated impedances. The unknown and calibrated impedances are connected to arms 1 and 2, the source to the parallel port P and the detector to the series port S. This is a null method: when the deflection of the detector is zero, the unknown and calibrated impedances are equal if they were connected at equal distances from the junction.

17.12 Power dividers

We have already met one type of waveguide power divider: the directional coupler. Some of the T-junctions described in the previous section can also be used as power dividers, but only in a 1 : 1 ratio. We shall now describe a few other types.

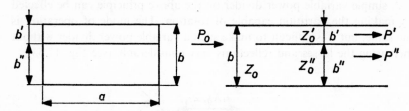

Figure 17.51

A very thin partition placed in the waveguide parallel to the wide sides (Fig. 17.51) so that the height b of the original guide is separated into b' and b'' divides the incident power in the ratio of the characteristic impedances Z_0' and Z_0'' of the two guides thus formed

$$\frac{Z_0'}{Z_0''} = \frac{b'}{b''} \qquad \frac{Z_0''}{Z_0} = \frac{b''}{b} \qquad (17.29)$$

$$b' + b'' = b \qquad Z_0' + Z_0'' = Z_0 \qquad (17.30)$$

If the two guides of reduced height are both terminated with a matched load, the input power P_0 is divided between them as follows

$$P' = \frac{Z_0''}{Z_0} = P_0 \qquad (17.31)$$

$$P'' = \frac{Z_0'}{Z_0} = P_0 \qquad (17.32)$$

The reflection is negligible if the partition (perpendicular to the electric lines of force) is thin enough. It is, of course, possible to increase the height of the two guides formed by the partition to the normal value by means of tapers (Fig. 17.52). The length of the taper is determined by trial and error.

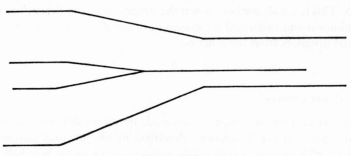

Figure 17.52

A simple variable power divider on the above principle can be effected by making the partition capable of rotation. The mode of operation is the same, but it is difficult to make such a variable power divider without increasing the losses and reflections very considerably (see Fig. 17.53).

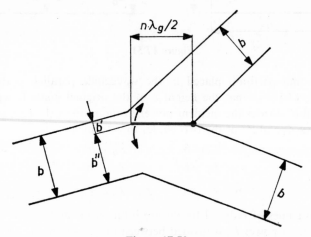

Figure 17.53

Finally, it is possible to make a waveguide power divider on the same principle as that of the coaxial type discussed in Section 12.8 (see Fig. 12.24). It should be remembered, however, that the T-junctions used are no longer reflection-free; special measures have to be taken to match the power divider in the desired frequency range, by means of diaphragms or the like.

Finally, we may observe that the power can be divided over a large number of guides by means of a modification of the first type mentioned above with a corresponding number of partitions.

17.13 Filters

We shall now briefly discuss a variety of filters, without going fully into design details.

High-pass filter

The high-pass filter makes use of the high-pass properties of the waveguide itself. All waves of frequency f such that $\lambda = c/f > 2a$, where a is the width of the guide, are exponentially attenuated. In order to separate high and low frequencies, therefore, we can make part of the

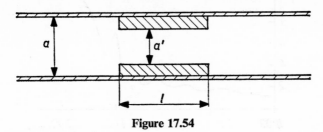

Figure 17.54

waveguide narrower than normal (Fig. 17.54). Only waves with $\lambda < 2a'$ are transmitted without attenuation. The length l should be chosen to give adequate attenuation at the frequency in question. The attenuation is given by

$$\alpha = 8 \cdot 69 \, \frac{2\pi}{2a'} \sqrt{1 - \left(\frac{2a'}{\lambda}\right)^2} \quad (\text{dB m}^{-1})$$

$$\text{for } \lambda \geqslant 1 \cdot 05\lambda_c \text{ and } \lambda_c = 2a' \quad (17.33)$$

Reflection of the desired frequencies by the discontinuities in width can be avoided by narrowing the waveguide with tapers (Fig. 17.55) or by making the length l an integral multiple of $\lambda_g/2$ in the middle of the

Figure 17.55

frequency band to be passed. Reflections then cancel out because the length l gives rise to a phase difference of 2π, while the reflection coefficients due to the changes in width from a to a' and back are of opposite sign (the one being due to a reduction in width, and the other to an increase).

If the frequency to be passed approaches the cut-off frequency of the narrow part of the waveguide too closely, the guide wavelength becomes very large, and it may be difficult or impossible to match the filter for the frequency in question (see also Fig. 17.56, in which the standing-wave ratio is plotted as a function of λ/λ_c).

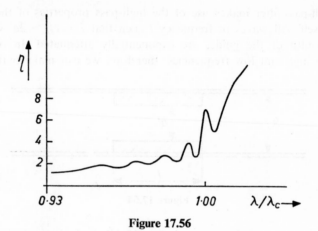

Figure 17.56

The literature also gives approximate expressions for the equivalent reactance of symmetrical changes in waveguide width from a to a'. The equivalent circuit is shown in Fig. 17.57. Passing from the wide guide to the narrow guide, we choose T as the reference plane. (For the opposite direction, another reference plane T' should be chosen in order to keep the expressions the same.) Now

$$\frac{X}{Z_0} \approx \frac{2a}{\lambda_g} \cdot 2\cdot33\alpha^2(1+1\cdot56\alpha^2)(1+6\cdot75\alpha^2 Q) \quad \text{for } \alpha \ll 1 \quad (17.34)$$

and

$$Q = 1 - \sqrt{1 - \left(\frac{2a}{3\lambda}\right)^2} \qquad \alpha = \frac{a'}{a} \qquad (17.35$$

We shall now give a sample calculation for a cut-off filter of this type. Let us suppose that waves of frequency $f_1 = 8500$ MHz should be attenuated by at least 30 dB, while $f_2 = 10\,000$ MHz should not be attenuated at all. The waveguide normally used for this frequency range has internal dimensions $a = 22\cdot86$ mm and $b = 10\cdot16$ mm. The free-space wavelengths corresponding to f_1 and f_2 are $\lambda_1 = 35\cdot27$ mm and $\lambda_2 = 29\cdot98$ mm respectively.

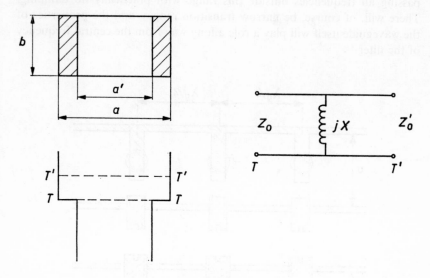

Figure 17.57

The cut-off wavelength of the narrow part, $\lambda_c = 2a'$, must thus lie between λ_1 and λ_2. If we choose λ_c too close to λ_1 with the object of avoiding reflections at λ_2 (Fig. 17.56), then the filter has to be very long if it is to attenuate λ_1 adequately. A number of widths are generally tried. We choose, say, $\lambda_c = 2a' = 33\cdot0$ mm. The attenuation rate at λ_1 is then

$$\alpha = 8\cdot69\,\frac{2\pi}{33}\,\sqrt{1 - \left(\frac{33\cdot0}{35\cdot27}\right)^2} = 0\cdot585 \text{ dB mm}^{-1}$$

In order to achieve an attenuation of 30 dB, therefore, we need a length l of at least

$$l = 30/0\cdot585 \approx 51 \text{ mm}$$

Now the waveguide wavelength for waves of frequency f_2 in the narrow part of the guide is given by

$$\lambda_{g2} = \lambda_2/\sqrt{1 - (\lambda_2/2a')^2} = 71\cdot6 \text{ mm}$$

We choose $l = \lambda_{g2}$ to minimize reflections due to the change in width; the attenuation at f_1 will then certainly be sufficient.

Band-stop filter

The band-stop filter reflects all frequencies within its bandwidth (so that the effective forward damping, for example, is 20 to 30 dB) while passing all frequencies outside this range with practically no damping. There will, of course, be narrow transition ranges, and the properties of the waveguide itself will play a role a long way from the central frequency of the filter.

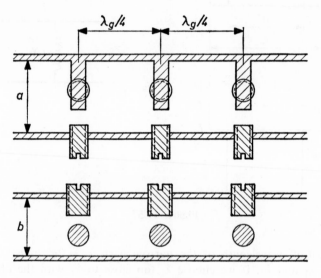

Figure 17.58

Fig. 17.58 shows two sections through one version of such a band-stop filter, which is built up of three parts. The fixed horizontal pin forms a series resonant circuit together with the adjustable screw in the opposite wall of the waveguide. These resonant circuits are tuned either to the middle of the stop band or to slightly different frequencies so as to increase the bandwidth somewhat. However, they are perpendicular to the electric field, and would not thus be very effective. Adjustable screws parallel to the field of the TE_{01} mode are therefore placed in the wall of the guide above the resonant circuits, in order to couple the latter with the field. The electric field of the principal mode is deformed by these screws, so that it is no longer perpendicular to the resonant circuits. The three sections are spaced at intervals of a quarter wavelength.

A similar band-stop filter can be constructed with cavity resonators as the resonant elements (Fig. 17.59). The length of the resonators, the size of the coupling hole and the number of sections must be calculated afresh for each particular case. The filter as a whole is less flexible than the version described above, in particular with regard to the strength of the coupling between the resonant circuits and the waveguide.

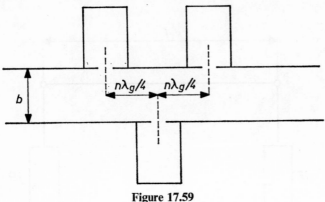

Figure 17.59

Band-pass filter

Fig. 17.60 shows an example of a band-pass filter in a waveguide, using three cavity resonators coupled by lengths of guide. The coupling can be made through circular holes in a wall or by means of an inductive diaphragm as discussed in Section 17.1. The cavity resonator is roughly tuned to the desired frequency by suitable choice of length l_1, l_2 and l_3. Fine adjustment can be done with the screw in the wide wall of the

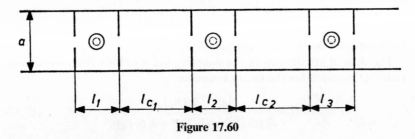

Figure 17.60

waveguide section in the middle of the cavity resonator (this has the effect of adding an extra capacitance or inductance as described in Section 17.1). The length of the cavity resonator for resonance would be half the guide wavelength if it were not for the coupling holes in the walls. However, the effective impedances of the coupling holes have an influence on the desired length of the cavity resonator. We shall now discuss this effect, without going into too much detail.

Fig. 17.61 illustrates a cavity resonator of length l, with a coupling-hole admittance jB on either side. The resonator has the same transverse dimensions as the waveguide in which it is placed. If Y_0 is the characteristic admittance of the waveguide, it follows from a simple calculation that the

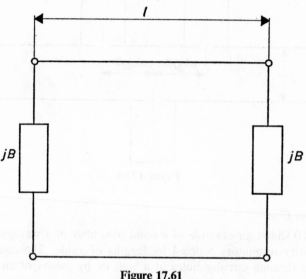

Figure 17.61

resonance condition for the circuit of Fig. 17.61 can be written

$$\tan \frac{2\pi l}{\lambda_{go}} = \frac{2B/Y_0}{(B/Y_0)^2 - 1} \approx \frac{2}{B/Y_0} \qquad (17.36)$$

for most cases occurring in practice $B/Y_0 \gg 1$.

The desired admittance of the coupling holes follows from the quality factor Q required for the cavity resonator

$$Q = \frac{\tan^{-1} 2/(B/Y_0)}{2 \sin^{-1} 2/\sqrt{(B/Y_0)^4 + 4(B/Y_0)^2}} \qquad (17.37)$$

Once B/Y_0 is known and the form of the coupling hole has been chosen (e.g. an inductive diaphragm), the dimensions of the hole can be calculated from data given in the literature (and partly in this book—see Section 17.1).

The waveguide sections joining the cavity resonators should, in principle, be a quarter wavelength long; however, these lengths should also be corrected for the impedance of the coupling holes. A reduced coupling-hole impedance of jB/Y_0 decreases the length between the two cavity resonators by a length l' given by $\tan(2\pi l'/\lambda_g) = -1/(B/Y_0)$. It follows from equation (17.36) that

$$l' = \frac{\lambda_g}{4} - \frac{l}{2}$$

The total length of the coupling piece is thus given by

$$l_{c1} = (2m+1)\frac{\lambda_g}{4} - l'_1 - l'_2 = (2m+1)\frac{\lambda_g}{4} - \left(\frac{\lambda_g}{4} - \frac{l_1}{2}\right) - \left(\frac{\lambda_g}{4} - \frac{l_2}{2}\right)$$

where m is an integer or zero.

It follows finally that

$$l_{c1} = \frac{l_1 + l_2}{2} - \frac{\lambda_g}{4} + m\frac{\lambda_g}{2} \qquad (17.38)$$

We generally choose l_c in the region of $3\lambda_g/4$, because $l_c \approx \lambda_g/4$ would cause too much interaction between neighbouring cavity resonators, while $l_c \approx 5\lambda_g/4$ would excessively increase the frequency dependence of the coupling pieces.

17.14 Flanges

We shall finish this chapter on waveguide components with a few comments about the flanges which provide the mechanical and electrical connection between successive sections of waveguide.

Many different types of flange are used and we cannot give a detailed treatment here, except to mention some of the demands they must satisfy.

The reflection due to a flanged joint must be negligible over the entire frequency range for which the waveguide is intended. Leakage of power from the waveguide through the joints must be negligible. The flanged joints must often be water-tight, or capable of being made water-tight (for example, with rubber sealing rings). This applies especially to set-ups for use in the open air, such as radar. The mechanical strength must be adequate, and the flanges must not be too large. Flanges should be easy to mount on the waveguide, and their coupling, to give a reflection-free joint, must be a quick, simple and fool-proof matter. In practice, most flanges are flat, being joined together by four or more screws or clips.

CAVITY RESONATORS

18 INTRODUCTION TO THE CAVITY RESONATOR

Resonators are widely used in oscillators, amplifiers, filters, wavemeters, etc. Resonant circuits built up of lumped capacitances and self-inductances, as used in low-frequency applications, cannot always be made in the same way in microwave circuits. As the frequency increases, a lumped capacitance incurs too much self-inductance of its own, often with accompanying high losses. A self-inductance will quickly acquire too high a capacitance between successive windings. However, with modern microminiaturization techniques, even in the microwave area, lumped elements, small compared with the wavelength, can now be used.

In the frequency range from about 100 to 3000 MHz, 'butterfly' resonators (Fig. 18.1) can conveniently be used. A number of the plates

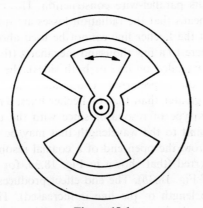

Figure 18.1

shown in the figure are stacked on top of one another; the ring forms the self-inductance, while the capacitance is produced between the stator (ring) and the rotor (middle part). Tuning ratios of 3 : 1 to 5 : 1 can be achieved in this way. These resonators are available in a number of different versions.

We have already discussed resonators in the form of a length of Lecher line or coaxial line (Section 11.2). A variety of types is illustrated in Fig. 18.2, together with the voltage distribution along the 'cavity' resonator for different lengths of the line section. The figure also shows the general circuit of a lumped reactance jX_t combined with a piece of parallel-wire line of length l and characteristic impedance Z_0. The resonance condition for this combination is

$$jZ_0 \tan \beta l = -jX_t \qquad (18.1)$$

Once the length of the line section is known, the resonance frequency can be calculated from this equation.

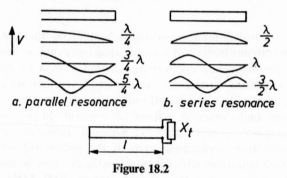

a. parallel resonance   b. series resonance

Figure 18.2

The Lecher line is simple, and is very suitable for applications with valves because of its parallel-wire construction. However, the fact that it is not shielded means that the radiation losses are quite high at higher frequencies, so that the Lecher line cannot be used above a few hundred megahertz. From here to a few thousand megahertz (the frequency range can sometimes be extended on one or both sides), the coaxial line, with its closed construction, is preferred. The quality factor of coaxial resonators is generally greater than that of Lecher lines. The tuning range is also wide with this type of resonator since with the usual TEM mode there is no lower limit to the wavelength that may be used. In order to prevent radiation from the open end of a coaxial resonator, closed constructions are preferred (that shown in Fig. 18.3a, for example, is then replaced by that of Fig. 18.3b). The end effect produces an extra capacitance (the effective length of the line is increased). The electron tubes used often terminate the line in this way, especially disc-seal triodes. The inner conductor may be insulated, so that it can carry d.c. voltages. Care should be taken to choose the transverse dimensions of the coaxial resonator so that higher modes cannot occur. (Higher modes would lead to power losses due to mode coupling if they had the same resonance frequency as the principal mode, and to a jump in the resonance frequency if the different modes had different resonance frequencies.)

Resonators made from waveguide sections resemble those made from lengths of parallel-wire line, although there are certain differences. For example, the tuning ratio is smaller: no propagation is possible below the cut-off frequency, and if only one or two modes are to be transmitted there is also an upper limit to the frequency. Moreover, the phase velocity depends on f/f_c. The quality factor is generally higher than that of coaxial resonators. Waveguide resonators are generally used at the higher micro-wave frequencies, above 3000 MHz.

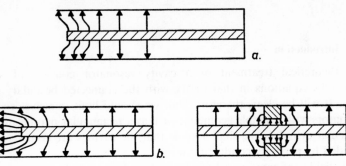

Figure 18.3

Finally, we have the more general cavity resonators, which may be regarded as sections of waveguide or coaxial line which are short-circuited on all sides (with the exception of the coupling holes, of course).

A cavity resonator is thus a space surrounded by walls of very high conductivity (which we shall assume to be perfect conductors). In prin-ciple, these cavities can have any shape, but only a few simple forms are used in practice. A very large number of resonant modes can generally occur in a 'box resonator' of this type. Attempts are often made to design the cavity resonator so that it will contain only one mode in a limited frequency range. Such resonators can be tuned by changing the shape or by introducing extra conductors or dielectric rods in appropriate places. These aids will normally be placed in regions of high field strength, so that they influence the tuning effectively. For the fine tuning, on the other hand, a less sensitive region should be chosen. The energy stored in a cavity resonator of this type is roughly proportional to the volume, while the wall losses are proportional to the surface area. The quality factor Q at a given frequency for a given mode will thus be proportional to the ratio of volume to surface area. If the resonator is fixed, then Q is proportional to the square root of the wavelength. Q is higher for higher modes, because the effective ratio of volume to surface area is higher for the higher modes. We shall now discuss a number of these cavity resonators in some detail.

19 SIMPLE CAVITY RESONATORS

19.1 Introduction

The theoretical treatment of a cavity resonator consists of solving Maxwell's equations in that cavity with the connected boundary conditions. It will be clear, therefore, that we should limit ourselves to cavity resonators with a simple geometry, such as a rectangular parallelepiped or a circular cylinder. We shall assume that the cavities are completely surrounded by perfectly conducting walls.

The field configuration follows from the solution of Maxwell's equations with the appropriate boundary conditions. The resonance frequencies of the various modes of vibration will appear as secondary conditions in the solution of the differential equations involved by separation of the variables (cf. the formula for Γ^2 found during the solution of the field equations for a rectangular waveguide). This thus gives us the form of the field and the resonance frequencies of the various modes in a cavity resonator with perfectly conducting walls. We assume that the form of the field and the resonance frequencies are not affected appreciably by the fact that the walls (copper, silver, etc.) actually have a finite conductivity. The quality factor Q_0 of a real cavity resonator is determined from the field configuration for the ideal resonator and the corresponding resonance frequency ω_0

$$Q_0 = \frac{2 \int\limits_V H^2 \, dV}{\delta \int\limits_S H^2 \, dS} \tag{19.1}$$

where δ is the penetration depth, V is the volume and S is the surface area of the resonator.

As we know, the quality factor is in a cavity resonator defined as

$$Q = \omega_0 W/P = \omega_0 \cdot \frac{\text{energy stored}}{\text{mean power dissipated}} \tag{19.2}$$

Now the amount of energy stored in the cavity is normally $(\frac{1}{2}\mu H^2 + \frac{1}{2}\varepsilon E^2)$ per unit volume. At resonance, all the energy is stored either in electric or magnetic form. If we choose the magnetic field at resonance in the cavity in equation (19.1) for H, we find

$$W = \tfrac{1}{2}\mu \int_V H^2 \, \mathrm{d}V \qquad (19.3)$$

Once we know H and the geometry for a particular cavity resonator we can find the full solution to this equation.

The power in an empty cavity resonator is all dissipated in the walls. We can therefore write

$$P = \tfrac{1}{2}R_S \int_S I_S^2 \, \mathrm{d}S \qquad (19.4)$$

where R_S is the surface resistance of the wall and I_S is the amplitude of the current density at the surface. Now we know from the general boundary conditions that the wall current is equal to the discontinuity in the tangential magnetic field at the wall. Since, however, we are using the magnetic field configuration of the ideal cavity with perfectly conducting walls in the first instance, the magnetic field has no tangential component in the wall. We may thus write

$$I_S = H_{\text{tang}} \qquad (19.5)$$

where H_{tang} is the tangential field strength at resonance in the cavity near the wall. Furthermore, if the walls are perfect conductors the normal component of the magnetic field must be continuous and zero. H_{tang} is therefore equal to the total field strength at the wall, so that

$$P = \tfrac{1}{2}R_S \int_S H^2 \, \mathrm{d}S \qquad (19.6)$$

The surface resistance R_S is determined by the frequency and the conductivity σ of the material, according to the equation

$$R_S = \sqrt{\frac{\pi \mu f}{\sigma}} \qquad (19.7)$$

We thus find

$$Q = \omega_0 \frac{W}{P} = \omega_0 \frac{\dfrac{1}{2}\mu \int_V H^2 \mathrm{d}V}{\dfrac{1}{2}\sqrt{\dfrac{\pi \mu f}{\sigma}} \int_S H^2 \, \mathrm{d}S} = 2\sqrt{\pi f_0 \mu \sigma}\; \frac{\int_V H^2 \mathrm{d}V}{\int_S H^2 \, \mathrm{d}S} = \frac{2}{\delta} \frac{\int_V H^2 \mathrm{d}V}{\int_S H^2 \, \mathrm{d}S}$$

$$(19.8)$$

This thus gives us the field configuration, ω_0 and Q_0 for the cavity resonator, but we must still couple the cavity with an external circuit through coupling holes, loops or the like. For wavemeters, for example, the coupling is weak, so that the quality factor of the loaded cavity Q_L will not differ much from the unloaded value Q_0 given above. Fig. 19.1 gives two examples of how the cavity resonator will be coupled: a as a

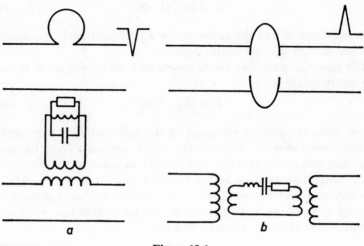

Figure 19.1

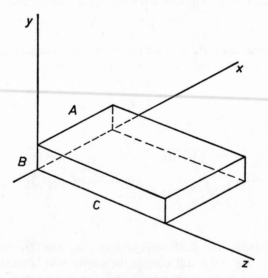

Figure 19.2

reaction resonator and b as a transmission resonator. In both cases, coupling is provided by a hole in the wall.

19.2 Rectangular parallelepiped

Fig. 19.2 shows a block placed with its longest edge C along the z axis, its shortest edge B along the y axis and the third edge A along the x axis. This cavity resonator does not contain any free charge ($\rho = 0$), and there are no conduction currents in the interior ($J = 0$). In order to solve Maxwell's equation for this situation (where there are no travelling waves), we again separate the field into TE and TM components, as we did with travelling waves in waveguides, so that the solution will be obtained in two parts: TE fields with $E_z = 0$ and TM fields with $H_z = 0$.

Firstly, we consider the case of $E_z = 0$. We again assume that the fields vary harmonically with time

$$E(x, y, z, t) = E_0(x, y, z)\, e^{j\omega t}$$
$$H(x, y, z, t) = H_0(x, y, z)\, e^{j\omega t} \tag{19.9}$$

Substitution in Maxwell's equations and elimination of E or H again gives the wave equations for the six components E_{0x} to H_{0z}. We solve for the E field in this way, and determine the H field from the solution found for the electric field together with Maxwell's equations. By an argument similar to that used for the rectangular waveguide (with E_{0x}, etc., instead of G_x, etc.) we find the equations for the E field to be

$$\frac{\partial^2 E_{0x}}{\partial x^2} + \frac{\partial^2 E_{0x}}{\partial y^2} + \frac{\partial^2 E_{0x}}{\partial z^2} = -\omega^2 \varepsilon\mu E_{0x} \tag{19.10}$$

The equations for E_{0y} and E_{0z} are found by cyclic permutation in x, y and z; however, $E_{0z} \equiv 0$ in this TE case. Furthermore

$$\frac{\partial E_{0x}}{\partial x} + \frac{\partial E_{0y}}{\partial y} + \frac{\partial E_{0z}}{\partial z} = 0$$

or

$$\frac{\partial E_{0x}}{\partial x} + \frac{\partial E_{0y}}{\partial y} = 0 \tag{19.11}$$

The boundary conditions may be written

$$E_{0y} = E_{0z} = 0 \quad \text{at } x = 0 \text{ and } x = A$$
$$E_{0x} = E_{0z} = 0 \quad \text{at } y = 0 \text{ and } y = B \tag{19.12}$$
$$E_{0x} = E_{0z} = 0 \quad \text{at } z = 0 \text{ and } z = C$$

Let us solve equation (19.10) by separating the variables. For this purpose, we write

$$E_{0x}(x, y, z) = X(x) . Y(y)Z(z) \tag{19.13}$$

Substitution in equation (19.10), after some manipulation, gives

$$\frac{1}{X}\frac{d^2X}{dx^2} + \frac{1}{Y}\frac{d^2Y}{dy^2} + \frac{1}{Z}\frac{d^2Z}{dz^2} = -\omega^2\varepsilon\mu \tag{19.14}$$

Since the right-hand side is constant, each of the three terms on the left-hand side must also be constant

$$\frac{1}{X}\frac{d^2X}{dx^2} = -P_1^2; \ \frac{1}{Y}\frac{d^2Y}{dy^2} = -Q_1^2; \ \frac{1}{Z}\frac{d^2Z}{dz^2} = -R_1^2 \tag{19.15}$$

where

$$P_1^2 + Q_1^2 + R_1^2 = \omega^2\mu\varepsilon \tag{19.16}$$

We can solve equation (19.15) as follows (similar to the previous case)

$$X = a_1 \sin P_1 x + b_1 \cos P_1 x \tag{19.17}$$
$$Y = c_1 \sin Q_1 y + d_1 \cos Q_1 y \tag{19.18}$$
$$Z = e_1 \sin R_1 z + f_1 \cos R_1 z \tag{19.19}$$

Substitution in equation (19.13) gives

$$E_{0x} = (a_1 \sin P_1 x + b_1 \cos P_1 x)(c_1 \sin Q_1 y + d_1 \cos Q_1 y) \times$$
$$\times (a_1 \sin R_1 z + f_1 \cos R_1 z) \tag{19.20}$$

and similarly

$$E_{0y} = (a_2 \sin P_2 x + b_2 \cos P_2 x)(c_2 \sin Q_2 y + d_2 \cos Q_2 y) \times$$
$$\times (e_2 \sin R_2 z + f_2 \cos R_2 z) \tag{19.21}$$

and

$$E_{0z} = 0 \tag{19.22}$$

Substitution of the boundary conditions in equations (19.20)–(19.22) gives

$$d_1 = 0 \qquad b_2 = 0 \qquad f_1 = 0 \qquad f_2 = 0$$
$$Q_1 = \frac{m\pi}{B} \qquad P_2 = \frac{l\pi}{A} \qquad R_1 = \frac{n\pi}{C} \qquad R_2 = \frac{n'\pi}{C} \tag{19.23}$$

where l, m, n and n' are zero or integers.

Substitution of equation (19.23) in the equations for E_{0x} and E_{0y} then gives

$$E_{0x} = (a_1 \sin P_1 x + b_1 \cos P_1 x)\left(c_1 \sin \frac{m\pi}{B} y\right)\left(e_1 \sin \frac{n\pi}{C} z\right)$$
$$E_{0y} = \left(a_2 \sin \frac{l\pi}{A} x\right)(c_2 \sin Q_2 y + d_2 \cos Q_2 y)\left(e_2 \sin \frac{n'\pi}{C} z\right)$$
$$E_{0z} = 0 \tag{19.24}$$

From equation (19.24), the condition (19.11) now becomes

$$(a_1 P_1 \cos P_1 x - b_1 P_1 \sin P_1 x)c_1 e_1 \sin \frac{m\pi}{B} y \sin \frac{n\pi}{C} z +$$
$$a_2 e_2 \sin \frac{l\pi}{A} x \sin \frac{n'\pi}{C} z(c_2 Q_2 \cos Q_2 y - d_2 Q_2 \sin Q_2 y) = 0 \tag{19.25}$$

This requirement holds for $x \leqslant A$, $y \leqslant B$, $z \leqslant C$.

Much as we did for the rectangular waveguide, we can simplify this somewhat by considering some situations in which the requirement must be satisfied. If it is to be met at $y = 0$, then

$$c_2 = 0 \qquad (19.26)$$

since $e_2 \neq 0$ and $a_2 \neq 0$; otherwise, E_{0y} would be zero.

If the requirement is to be met at $y = B$, then

$$Q_2 = \frac{m'\pi}{B} \qquad (19.27)$$

since $a_2 \neq 0$, $e_2 \neq 0$ and $d_2 \neq 0$; otherwise, E_{0y} would be zero.

If the requirement is to be met at $x = 0$, then

$$a_1 = 0 \qquad (19.28)$$

since $c_1 \neq 0$ and $e_1 \neq 0$; otherwise, E_{0x} would be zero.

Similarly, it follows from the requirement at $x = A$ that

$$P_l = \frac{l'\pi}{A} \qquad (19.29)$$

Substitution of equations (19.26)–(19.29) in (19.25) gives the remaining requirement for $x \leqslant A$, $y \leqslant B$ and $z \leqslant C$

$$a_2 e_2 d_2 \frac{m'\pi}{B} \sin \frac{l\pi}{A} x \sin \frac{m'\pi}{B} y \sin \frac{n'\pi}{C} z +$$

$$b_1 c_1 e_1 \frac{l'\pi}{A} \sin \frac{l'\pi}{A} x \sin \frac{m\pi}{B} y \sin \frac{n\pi}{C} z = 0 \quad (19.30)$$

This can only be satisfied in the (x, y, z) region in question if

$$l' = l, \qquad m' = m, \qquad n' = n \qquad (19.31)$$

and if

$$\frac{m\pi}{B} a_2 d_2 e_2 + \frac{l\pi}{A} b_1 c_1 e_1 = 0 \qquad (19.32)$$

We now introduce a few new variables

$$k_1 = \frac{l\pi}{A}, \qquad k_2 = \frac{m\pi}{B}, \qquad k_3 = \frac{n\pi}{C},$$

$$\qquad (19.33)$$

$$k = \sqrt{k_1^2 + k_2^2 + k_3^2} = \omega\sqrt{\varepsilon\mu} = \frac{\omega}{c} = \frac{2\pi}{\lambda}$$

$$b_1 c_1 e_1 = T_0 \qquad (19.34)$$

as general amplitude factor.

Finally, substituting the boundary conditions and the further requirements on the electric field we find

$$E_{0x} = T_0 \cos k_1 x \sin k_2 y \sin k_3 z$$

$$E_{0y} = -T_0 \frac{k_1}{k_2} \sin k_1 x \cos k_2 y \sin k_3 z \tag{19.35}$$

$$E_{0z} = 0$$

The expressions for the magnetic field can be found from the above with the aid of Maxwell's equations

$$H_{0x} = \frac{1}{j\omega\mu} \frac{\partial E_{0y}}{\partial z} = j \sqrt{\frac{\varepsilon}{\mu}} T_0 \frac{k_1 k_3}{k_2 k} \sin k_1 x \cos k_2 y \cos k_3 z$$

$$H_{0y} = -\frac{1}{j\omega\mu} \frac{\partial E_{0x}}{\partial z} = j \sqrt{\frac{\varepsilon}{\mu}} T_0 \frac{k_2 k_3}{k_2 k} \cos k_1 x \sin k_2 y \cos k_3 z \tag{19.36}$$

$$H_{0z} = -\frac{1}{j\omega\mu} \left[\frac{\partial E_{0x}}{\partial x} - \frac{\partial E_{0x}}{\partial y} \right] = -j \sqrt{\frac{\varepsilon}{\mu}} T_0 \frac{k_1^2 + k_2^2}{k_2 k} \cos k_1 x \cos k_2 y \sin k_3 z$$

By introducing a new constant S_0, defined by

$$T_0 = -\sqrt{\frac{\mu}{\varepsilon}} \frac{k_2}{k} S_0 \tag{19.37}$$

(N.B. k_2/k is constant), we can write the expressions for the fields in a more symmetrical way

$$E_x = -\sqrt{\frac{\mu}{\varepsilon}} S_0 \frac{k_2}{k} \cos k_1 x . \sin k_2 y . \sin k_3 z . e^{j\omega t}$$

$$E_y = \sqrt{\frac{\mu}{\varepsilon}} S_0 \frac{k_1}{k} \sin k_1 x . \cos k_2 y . \sin k_3 z . e^{j\omega t} \tag{19.38}$$

$$E_z = 0$$

$$H_x = -jS_0 \frac{k_1 k_3}{k^2} \sin k_1 x . \cos k_2 y . \cos k_3 z . e^{j\omega t}$$

$$H_y = -jS_0 \frac{k_2 k_3}{k^2} \cos k_1 x . \sin k_2 y . \cos k_3 z . e^{j\omega t} \tag{19.39}$$

$$H_z = jS_0 \frac{k_1^2 + k_2^2}{k^2} \cos k_1 x . \cos k_2 y . \sin k_3 z . e^{j\omega t}$$

Equations (19.38) and (19.39) thus represent the TE electromagnetic field in a rectangular box.

We may remind the reader that we must have

$$n \neq 0$$

and

$$m \text{ or } l \neq 0 \tag{19.40}$$

since the field would otherwise be equal to zero, and

$$k^2 = k_1^2 + k_2^2 + k_3^2 = \left(\frac{2\pi}{\lambda} \right)^2 \tag{19.41}$$

In the above l indicates the number of half waves along the x axis; m indicates the number of half waves along the y axis; and n indicates the number of half waves along the z axis. We choose these numbers positive or zero.

The expressions for the TM field can be derived in much the same way as that described above for the TE field. We shall only give the result here

$$E_x = \sqrt{\frac{\mu}{\varepsilon}}\, S_0 \frac{k_1 k_3}{k^2} \cos k_1 x.\, \sin k_2 y.\, \sin k_3 z.\, e^{j\omega t}$$

$$E_y = \sqrt{\frac{\mu}{\varepsilon}}\, S_0 \frac{k_2 k_3}{k^2} \sin k_1 x.\, \cos k_2 y.\, \sin k_3 z.\, e^{j\omega t} \qquad (19.42)$$

$$E_z = -\sqrt{\frac{\mu}{\varepsilon}}\, S_0 \frac{k_1^2 + k_2^2}{k^2} \sin k_1 x.\, \sin k_2 y.\, \cos k_3 z.\, e^{j\omega t}$$

$$H_x = -jS_0 \frac{k_2}{k} \sin k_1 x.\, \cos k_2 y.\, \cos k_3 z.\, e^{j\omega t}$$

$$H_y = jS_0 \frac{k_1}{k} \cos k_1 x.\, \sin k_2 y.\, \cos k_3 z.\, e^{j\omega t} \qquad (19.43)$$

$$H_z = 0$$

We choose $l \neq 0$ and $m \neq 0$ in the above equations for the TM field, since the whole field would otherwise be zero. Of course, equation (19.41) also applies here.

For each combination (l, m, n), there is one possible TE_{lmn} mode and one possible TM_{lmn} mode in the box. It will be clear that many different resonances are possible. The expression for the resonance frequency (eigen frequency) of the various modes is implicit in the above derivation of the electromagnetic field pattern (equations 19.16 and 19.41)

$$\left(\frac{2\pi}{\lambda}\right)^2 = \left(\frac{l\pi}{A}\right)^2 + \left(\frac{m\pi}{B}\right)^2 + \left(\frac{n\pi}{C}\right)^2$$

the resonance wavelength may thus be written

$$\lambda_0 = \frac{2}{\sqrt{\left(\dfrac{l}{A}\right)^2 + \left(\dfrac{m}{B}\right)^2 + \left(\dfrac{n}{C}\right)^2}} \qquad (19.44)$$

This expression holds for TE_{lmn} modes as well as for TM_{lmn} modes.

We could use the field expressions of equations (19.38) and (19.39), or (19.42) and (19.43), together with that for the resonance frequency (19.44), to determine the quality factor for a rectangular box resonator with walls of finite conductivity, along the lines described in Section 19.1. We shall not give the calculations in full here. The field variables given above are substituted in equation (19.8) but the procedure is rather cumbersome to manipulate. The result is generally written as an expression for $Q\delta/\lambda$, since this depends only on the mode and the shape (and size) of the cavity resonator.

As an example, we shall only give the expression for the TE mode with both l and m positive. (The expression becomes somewhat different if l or m is zero, and that for the TM mode is different again. These results can be found in the literature if desired.)

$$Q\frac{\delta}{\lambda} = \frac{ABC}{4}\frac{(p_1^2+p_2^2)(p_1^2+p_2^2+p_3^2)^{3/2}}{AC[p_1^2p_3^2+(p_1^2+p_2^2)^2]+} \qquad (19.45)$$
$$BC[p_2^2p_3^2+(p_1^2+p_2^2)^2]+ABp_3^2[p_1^2+p_2^2]$$

where

$$p_1 = \frac{k_1}{\pi}, \quad p_2 = \frac{k_2}{\pi} \quad \text{and} \quad p_3 = \frac{k_3}{\pi} \qquad (19.46)$$

19.3 Circular cylinders

For the cylindrical cavity resonator, we choose cylindrical co-ordinates r, θ and z as shown in Fig. 19.3. The problem is again to solve Maxwell's equations with their boundary conditions, this time for the cylindrical space in question. We assume that this space contains no electric charge or current: $\rho = 0$ and $J = 0$. Moreover, we again assume that the field is harmonic in time, according to $e^{j\omega t}$.

Written in cylindrical co-ordinates, Maxwell's equations become

$$\frac{1}{r}\frac{\partial E_z}{\partial \theta} - \frac{\partial E_\theta}{\partial z} = -j\omega\mu H_r$$

$$\frac{\partial E_r}{\partial z} - \frac{\partial E_z}{\partial r} = -j\omega\mu H_\theta \qquad (19.47)$$

$$\frac{1}{r}\frac{\partial}{\partial r}(rE_\theta) - \frac{1}{r}\frac{\partial E_r}{\partial \theta} = -j\omega\mu H_z$$

$$\frac{1}{r}\frac{\partial H_z}{\partial \theta} - \frac{\partial H_\theta}{\partial z} = j\omega\varepsilon E_r$$

$$\frac{\partial H_r}{\partial z} - \frac{\partial H_z}{\partial r} = j\omega\varepsilon E_\theta \qquad (19.48)$$

$$\frac{1}{r}\frac{\partial}{\partial r}(rH_\theta) - \frac{1}{r}\frac{\partial H_r}{\partial \theta} = j\omega\varepsilon E_z$$

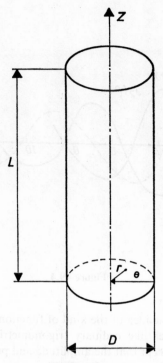

Figure 19.3

We also have the divergence equations

$$\frac{\partial}{\partial r}(rH_r) + \frac{\partial H_\theta}{\partial \theta} + r\frac{\partial H_z}{\partial z} = 0$$

$$\frac{\partial}{\partial r}(rE_r) + \frac{\partial E_\theta}{\partial \theta} + r\frac{\partial E_z}{\partial z} = 0$$

(19.49)

The wave equations for the six components of the electromagnetic field can again be derived from these forms of Maxwell's laws. It may be noted that these wave equations no longer have the same form for all six co-ordinates, as was the case with cartesian co-ordinates. The wave equations are obtained with the aid of Hertz vectors and separation into TE and TM components. This procedure would take us too long to describe in detail here.

Once the wave equations have been obtained, they can again be solved by separation of the variables. This gives rise to Bessel's equation with Bessel functions $J_l(r)$ as the solution. $J_l(r)$ represents a Bessel function of the first sort, order l and argument r. Fig. 19.4 shows the approximate form of some low-order Bessel functions for low values of the argument,

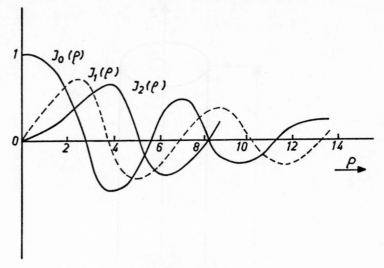

Figure 19.4

to give the reader some idea of the kind of functions involved. As may be seen, they look rather like ordinary trigonometric functions, having a periodic character, except that the amplitude and period are not constant. The derivative of the Bessel function $J_l(r)$ is written $J_l'(r)$, where the prime refers to differentiation with respect to the argument, i.e.

$$J_l'(\rho) = \frac{d}{d\rho} J_l(\rho) \tag{19.50}$$

We shall not go into details of the derivation here, but merely give the results of solving the cylindrical wave equations.

The field in a cylindrical cavity for TM waves may be written

$$E_r = -\sqrt{\frac{\mu}{\varepsilon}} \, S_0 \, \frac{k_3}{k} \, J_l'(k_1 r) . \cos l\theta . \sin k_3 z . e^{j\omega t}$$

$$E_\theta = \sqrt{\frac{\mu}{\varepsilon}} \, S_0 \, \frac{k_3}{k} \, \frac{J_l(k_1 r)}{k_1 r} \sin l\theta . \sin k_3 z . e^{j\omega t} \tag{19.51}$$

$$E_z = \sqrt{\frac{\mu}{\varepsilon}} \, S_0 \, \frac{k_1}{k} \, J_l(k_1 r) . \cos l\theta . \cos k_3 z . e^{j\omega t}$$

$$H_r = -jS_0 l \, \frac{J_l(k_1 r)}{k_1 r} \sin l\theta . \cos k_3 z . e^{j\omega t}$$

$$H_\theta = -jS_0 J_l'(k_1 r) . \cos l\theta . \cos k_3 z . e^{j\omega t} \tag{19.52}$$

$$H_z' = 0$$

where

$$k_1 = \frac{2p_{lm}}{D} \qquad k_3 = \frac{n\pi}{L} \qquad k^2 = k_1^2 + k_3^2 = \left(\frac{2\pi}{\lambda}\right)^2 \tag{19.53}$$

We have, moreover, assumed $m > 0$, while l, m and n have the same significance as in the rectangular parallelepiped. Further, p_{lm} is the mth zero of $J_l(\rho)$ (see Fig. 19.4).

The expressions for the TE electromagnetic field in the cylindrical cavity resonator are

$$E_r = -\sqrt{\frac{\mu}{\varepsilon}} \, S_0 l \, \frac{J_l(k_1 r)}{k_1 r} \sin l\theta. \sin k_3 z. \, e^{j\omega t}$$

$$E_\theta = -\sqrt{\frac{\mu}{\varepsilon}} \, S_0 J_l'(k_1 r). \cos l\theta. \sin k_3 z. \, e^{j\omega t} \tag{19.54}$$

$$E_z = 0$$

$$H_r = -jS_0 \frac{k_3}{k} J_l'(k_1 r). \cos l\theta. \cos k_3 z. \, e^{j\omega t}$$

$$H_\theta = jS_0 l \frac{k_3}{k} \frac{J_l(k_1 r)}{k_1 r} \sin l\theta. \cos k_3 z. \, e^{j\omega t} \tag{19.55}$$

$$H_z = -jS_0 \frac{k_1}{k} J_l(k_1 r). \cos l\theta. \sin k_3 z. \, e^{j\omega t}$$

The values of m and n should be positive for TE waves.

The other symbols have the same significance as for TM waves, apart from p_{lm}, which is now the mth zero of $J_l'(\rho)$. In both cases, S_0 is an arbitrary amplitude factor.

We need to know the roots p_{lm} of $J_l(\rho) = 0$ and $J_l'(\rho) = 0$ to determine the resonance frequencies (and quality factors) of the various TM and TE modes in the resonator. These roots, i.e. the zeros of $J_l(\rho)$ and $J_l'(\rho)$, are tabulated in the literature. A few of these roots are given in Table 19.1. The first and third columns contain the values of l, m and n in question, and the second and fourth columns the corresponding values of p_{lm}.

The equation for the resonance frequency is derived from the boundary conditions in the same way as described above for the rectangular parallelepiped. For TM waves, we have $E_\theta = 0$ for $r = D/2$, or $J_l(k_1 D/2) = 0$; while for TE waves we also have $E_\theta = 0$ for $r = D/2$, which in this case gives $J_l'(k_1 D/2) = 0$. We have called the roots of this equation p_{lm}, and may therefore write

$$k_1 \frac{D}{2} = p_{lm} \quad \text{or} \quad k_1 = \frac{2p_{lm}}{D}$$

so that

$$k^2 = \left(\frac{2\pi}{\lambda}\right)^2 = k_1^2 + k_3^2 = \left(\frac{2p_{lm}}{D}\right)^2 + \left(\frac{n\pi}{L}\right)^2$$

Table 19.1

TE_{lmn}	$J_l'(p_{lm}) = 0$	TM_{lmn}	$J_l(p_{lm}) = 0$
11 n	1·841	01 n	2·465
21 n	3·054	11 n	3·832
01 n	3·832	21 n	5·136
31 n	4·201	02 n	5·520
41 n	5·318	31 n	6·380
12 n	5·332	12 n	7·016
51 n	6·415	41 n	7·588
22 n	6·706	22 n	8·417
02 n	7·016	03 n	8·654
61 n	7·501	51 n	8·772
32 n	8·016	32 n	9·761
13 n	8·536	61 n	9·936
71 n	8·578	13 n	10·174

The expression for the resonance frequency thus becomes

$$\lambda_{res} = \frac{2}{\sqrt{\left(\dfrac{2p_{lm}}{\pi D}\right)^2 + \left(\dfrac{n}{L}\right)^2}} \tag{19.56}$$

for both TE_{lmn} and TM_{lmn} modes, where p_{lm} is the mth root of $J_l(\rho) = 0$ for the TM waves, and of $J_l'(\rho) = 0$ for the TE waves.

Instead of the resonance wavelength in equation (19.56) we can, of course, calculate the resonance frequency directly from

$$(f_{res}D)^2 = \left(\frac{cp_{lm}}{\pi}\right)^2 + \left(\frac{cn}{2}\right)^2 \left(\frac{D}{L}\right)^2 \tag{19.57}$$

where c is the velocity of light.

If we plot $(f_{res}D)^2$ from equation (19.57) against $(D/L)^2$, we obtain a straight line. A graph of this sort is called a *mode diagram* (Fig. 19.5). For any value of D/L, we can read off directly from the graph which modes will be of importance in a limited frequency range. The point of intersection with the vertical axis is $(cp_{lm}/\pi)^2$.

The following modes intersect with the vertical axis successively

$$TE_{11n}, TM_{01n}, TE_{21n}, TE_{01n} \text{ and } TM_{11n}, TE_{31n}, \text{ etc.}$$

(See also Table 19.1.) One such intersection with the vertical axis thus corresponds to a whole succession of modes

$$TE_{111}, TE_{112}, TE_{113}, \text{ etc., } TM_{011}, TM_{012}, TM_{013}, \text{ etc.}$$

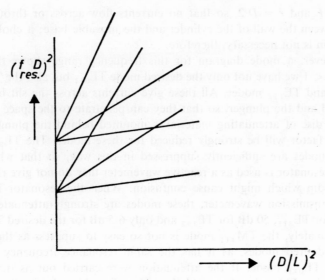

Figure 19.5

Finally, we shall describe an example of a cylindrical cavity resonator for use as a wavemeter in the 30 mm wavelength range (e.g. 8600 to 9600 MHz). It would be convenient if we could use the TE_{011} mode for this purpose, because then

$$f_{res} = \frac{1}{D} \sqrt{\left(\frac{c p_{01}}{\pi}\right)^2 + \left(\frac{cD}{2L}\right)^2} \tag{19.58}$$

so that the cavity resonator could be tuned by varying L. This can be done by using an adjustable short-circuit plunger for the base of the cavity, the position of the base being read off on a micrometer screw.

The electromagnetic field for the TE_{011} mode is given by

$$E_r = 0$$

$$E_\theta = -\sqrt{\frac{\mu}{\varepsilon}} S_0 J_0'\left(\frac{2p_{01}}{D} r\right) \sin \frac{\pi}{L} z \cdot e^{j\omega t} \tag{19.59}$$

$$E_z = 0$$

$$H_r = -jS_0 \frac{\lambda_{res}}{2L} J_0'\left(\frac{2p_{01}}{D} r\right) \cos \frac{\pi}{L} z \cdot e^{j\omega t}$$

$$H_\theta = 0 \tag{19.60}$$

$$H_z = -jS_0 \frac{p_{01}}{D} \frac{\lambda_{res}}{\pi} J_0\left(\frac{2p_{01}}{D} r\right) \sin \frac{\pi}{L} z \cdot e^{j\omega t}$$

Apart from the components which are identically equal to zero for this mode (E_r, E_z and H_θ), the other components (E_θ, H_r and H_z) are zero at $z = L$ and $r = D/2$, so that no currents flow across or through the slit between the wall of the cylinder and the movable base; a choke construction is not necessary, therefore.

However, a mode diagram for this frequency range shows that for $(D/L)^2 < 1$ we have not only the desired mode TE_{011} but also the TM_{111}, TE_{311} and TE_{112} modes. All these give currents across the slit between the wall and the plunger, so that they can penetrate to the space behind it. If a disc of attenuating material is mounted behind the plunger, the quality factor will be strongly reduced for these modes. The TE_{311} and TE_{112} modes are sufficiently suppressed in this way, so that when the cavity resonator is used as a reactive wavemeter they do not give rise to a visible dip which might cause confusion. When the resonator is used as a transmission wavemeter, these modes are strongly attenuated (e.g. 26 dB for TE_{311}, 50 dB for TE_{112} and only 6·5 dB for the desired TE_{011}). Unfortunately, the TM_{111} mode is not so easy to suppress as the other two undesired modes, as it has the same resonance frequency as the desired TE_{011} mode. If the attenuation were carried out as described above, this TM_{111} mode would thus strongly reduce the quality factor for the desired mode, and care must be taken to prevent it from being

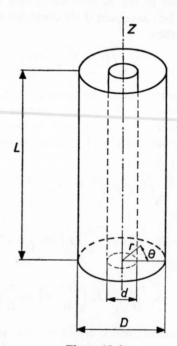

Figure 19.6

generated, by a suitable position of the coupling hole. Moreover, measures must be taken to prevent the electromagnetic vibration changing from the TE_{011} to the TM_{111} mode. Such measures include fitting the bottom and top of the cylindrical cavity resonator very accurately at right angles to the axis of the cylinder.

19.4 Coaxial cavity resonators

A coaxial cavity resonator is illustrated in Fig. 19.6 The mathematical treatment is similar to that of Section 19.3, except that the mathematics involved is even further beyond the scope of this book. Bessel functions of the second sort (Neumann functions) occur in the solutions of the differential equations; these are also tabulated in the relevant literature. The expressions for the electromagnetic field have the same form as those for the circular cylinder, when the following substitutions are made

$$Z_l(k_1 r) \text{ and } Z_l'(k_1 r) \text{ replace } J_l(k_1 r) \text{ and } J_l'(k_1 r)$$

where

$$Z_l(k_1 r) = J_l(k_1 r) - A Y_l(k_1 r) \tag{19.61}$$

$$Z_l'(k_1 r) = J_l'(k_1 r) - A Y_l'(k_1 r) \tag{19.62}$$

$Y_l(\rho)$ in the above equations is the Neumann function. It should be noted that $m > 0$ and $n > 0$ for TE modes, while $m > 0$ for TM modes.

The factor A in equations (19.61) and (19.62) is defined as

$$A = \frac{J_l(p_{lm})}{Y_l(p_{lm})} \tag{19.63}$$

for TM modes, where p_{lm} is the mth zero of

$$\{J_l(\eta\rho) Y_l(\rho) - J_l(\rho) Y_l(\eta\rho)\} \tag{19.64}$$

and

$$\eta = d/D. \tag{19.65}$$

For TE modes

$$A = \frac{J_l'(p_{lm})}{Y_l'(p_{lm})} \tag{19.66}$$

where p_{lm} is the mth zero of

$$\{J_l'(\eta\rho) Y_l(\rho) - J_l'(\rho) Y_l'(\eta\rho)\} \tag{19.67}$$

and η is given by equation (19.65).

20 EQUIVALENT CIRCUITS AND COUPLING COEFFICIENTS OF CAVITY RESONATORS

Consider a cavity resonator of arbitrary form, with resonance frequency $f_0 = \omega_0/2\pi$ and unloaded quality factor Q_0, coupled with a source of internal impedance Z_0 via a waveguide (say, rectangular) of characteristic impedance Z_0 (see Fig. 20.1).

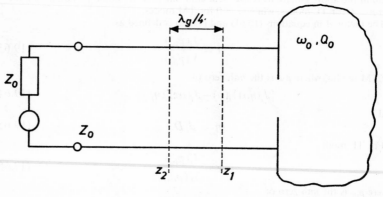

Figure 20.1

An ideal cavity resonator (i.e. without coupling holes and with entirely closed walls) can, when regarded as a resonant circuit, be represented by an lcr parallel circuit or an LCR series circuit, depending on the position of the reference plane in the coupling line from which the cavity resonator is seen. If the impedance at $z = z_1$, looking to the right, is minimum at resonance, then the resonance circuit and the length of input line to the right of $z = z_1$ are represented by a series circuit. At $z = z_2$, a quarter wavelength to the left, the impedance to the right is maximum (see the Smith diagram), and in this case everything to the right of $z = z_2$ can be represented by a parallel circuit.

244

For the series circuit, i.e. at $z = z_1$, we may write

$$\omega_0 = \frac{1}{\sqrt{LC}} \quad \text{and} \quad Q_0 = \frac{\omega_0 L}{R} \tag{20.1}$$

The load to the left of $z = z_1$ is Z_0 (waveguide with matched termination), so that the loaded quality factor of the whole circuit can be written (Fig. 20.2)

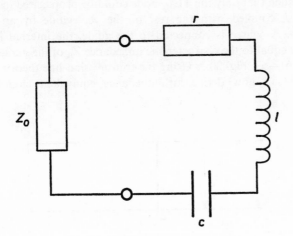

Figure 20.2

$$Q_L = \frac{\omega_0 L}{R + Z_0} \tag{20.2}$$

or alternatively

$$\frac{1}{Q_L} = \frac{R + Z_0}{\omega_0 L} = \frac{R}{\omega_0 L}\left(1 + \frac{Z_0}{R}\right) = \frac{1}{Q_0} + \frac{1}{Q_0}\frac{Z_0}{R} = \frac{1}{Q_0} + \frac{\beta}{Q_0} \tag{20.3}$$

where $\beta = Z_0/R$ is called the coupling coefficient. If we let

$$Q_{\text{ext}} = \frac{Q_0}{\beta} \tag{20.4}$$

be the external quality factor, we can write

$$\frac{1}{Q_L} = \frac{1}{Q_0} + \frac{1}{Q_{\text{ext}}} \tag{20.5}$$

If $\beta = 1$ or $Z_0 = R$, then

$$Q_0 = Q_{\text{ext}} = 2Q_L \tag{20.6}$$

When equation (20.6) is satisfied, the cavity resonator is said to be critically coupled to the waveguide; if $\beta < 1$, the cavity resonator is said to be undercoupled; while if $\beta > 1$, it is overcoupled. The magnitude of β can, of course, be controlled by varying the size of the coupling hole.

Finally, we shall consider the cavity resonator and coupling hole in somewhat more detail, and also take the impedance of the coupling hole into consideration. Let us consider a cavity resonator consisting of a short-circuited piece of waveguide (the transverse dimensions of which are such that only the TE_{01} mode could be propagated in the guide) of length l, coupled with the rest of the waveguide by an inductive diaphragm. A source is connected to the guide, the internal impedance of which is equal to the characteristic impedance Z_0 of the guide (matched termination)—see Fig. 20.3. Using the transmission-line theory developed above, we can now derive an impedance equivalent circuit for this situation.

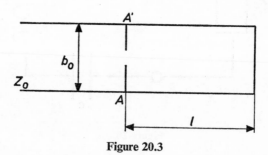

Figure 20.3

Suppose that the diaphragm has a susceptance $-jb$ or, reduced with respect to the characteristic admittance Y_0 of the waveguide, $-jb/Y_0$. This quantity is equal to the equivalent shunt admittance of the diaphragm in the plane of the diaphragm, A–A'. We assume that the diaphragm is of negligible thickness and that the losses in the short-circuit plane are negligible compared with the wall losses of the waveguide.

The input impedance of a piece of short-circuited transmission line of length l, characteristic impedance Z_0 and attenuation constant α can be written

$$Z_S = Z_0 \tanh (\alpha + j\beta)l = Z_0 \tanh \left(\alpha + j \frac{2\pi}{\lambda_g} \right) l \qquad (20.7)$$

Now l is not too large (not more than a few times $\lambda_g/2$), and we make the waveguide of a good conductor, such as copper or silver, so that α will be low. If $\alpha l \ll 1$, we may write $\tanh \alpha l \approx \alpha l$, so that

$$Z_S \approx Z_0 \frac{\alpha l + \tanh j\beta l}{1 + \alpha l \tanh j\beta l} = Z_0 \frac{\alpha l + j \tan \beta l}{1 + j\alpha l \tan \beta l} \qquad (20.8)$$

Furthermore, at resonance the length l will not be exactly equal to a whole number of half wavelengths. The coupling with the waveguide will give rise to a slight divergence, so that we should write

$$l \approx n\lambda_g/2 \tag{20.9}$$

where n is an integer. It follows that

$$\beta l \approx n\pi \tag{20.10}$$

and

$$|\tan \beta l| \ll l \tag{20.11}$$

If both $\alpha l \ll 1$ and $|\tan \beta l| \ll 1$, we may write

$$Z_S \approx Z_0(\alpha l + j \tan \beta l) \tag{20.12}$$

Let the resonance frequency for $l = \lambda_g/2$, i.e. for $\tan \beta l = 0$, ω_0 and the difference in resonance frequency owing to $l \approx \lambda_g/2$ (because of the diaphragm)

$$\Delta\omega = \delta = \omega - \omega_0 \tag{20.13}$$

The difference in length can similarly be written

$$l = \frac{\lambda_g}{2} - \Delta\frac{\lambda_g}{2} \tag{20.14}$$

If the influence of the coupling hole on the effective length and hence on the resonance wavelength of the cavity resonator is not too large, we may thus write

$$\begin{aligned}
\tan \beta l &= \tan \frac{2\pi l}{\lambda_g} = \tan \frac{2\pi}{\lambda_g}\left(\frac{\lambda_g}{2} - \Delta\frac{\lambda_g}{2}\right) = \tan\left(\pi - \pi\frac{\Delta\lambda_g}{\lambda_g}\right) \\
&= -\tan\frac{\pi\Delta\lambda_g}{\lambda_g} = -\pi\frac{\Delta\lambda_g}{\lambda_g} = -\frac{\pi}{\lambda_g}\left(\frac{\lambda_g}{\lambda}\right)^3 \Delta\lambda = \frac{\pi}{\lambda_g}\left(\frac{\lambda_g}{\lambda}\right)^3 \lambda\frac{\Delta\omega}{\omega_0} \\
&= \frac{\lambda_g^2}{\lambda^2}\frac{\pi\delta}{\omega_0}
\end{aligned} \tag{20.15}$$

We then find the approximation for the input impedance of the cavity resonator to the right of AA'

$$Z_S \approx Z_0\left(\alpha l + j\frac{\lambda_g^2}{\lambda^2}\frac{\pi\delta}{\omega_0}\right) \tag{20.16}$$

The expression (20.16) bears a close resemblance to that for the impedance of a series LCR circuit near the resonance frequency

$$Z_S' = R + j\left(\omega L - \frac{1}{\omega C}\right) = R + j2L\delta \tag{20.17}$$

where

$$\omega = \omega_0 + \Delta\omega = \omega_0 + \delta \quad \text{and} \quad \omega_0^2 = 1/LC$$

By analogy with this well-known resonance circuit we can derive the equivalent circuit of the cavity resonator by comparison of equations (20.16) and (20.17). This gives

$$R = Z_0 \alpha l$$

$$L = \tfrac{1}{2} Z_0 \frac{\lambda_g^2}{\lambda^2} \frac{\pi}{\omega_0}$$

(20.18)

while the susceptance of the diaphragm may be written

$$-jb = -j\frac{1}{\omega L_i}$$

(20.19)

The equivalent circuit for the cavity resonator and diaphragm, as seen from the plane of the diaphragm, is thus as shown in Fig. 20.4. Remembering that the matched source is connected in parallel with the above, we find the overall equivalent circuit for the system near resonance as shown in Fig. 20.5. This can, of course, be transformed into a series

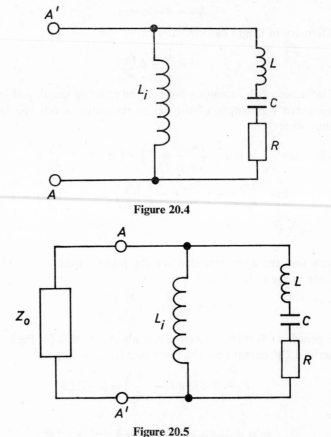

Figure 20.4

Figure 20.5

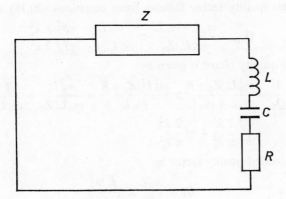

Figure 20.6

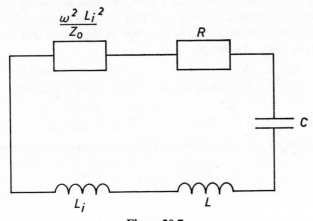

Figure 20.7

equivalent circuit (Fig. 20.6) for source, diaphragm and cavity resonator. The parallel combination of L_i and Z_0 has an impedance

$$Z = \frac{j\omega L_i Z_0}{j\omega L_i + Z_0} = j\omega L_i \frac{1}{1 + \omega^2 L_i^2/Z_0^2} + \frac{\omega^2 L_i^2}{Z_0} \frac{1}{1 + \omega^2 L_i^2/Z_0^2} \quad (20.20)$$

Cavity resonators with high values of Q and small coupling holes have $\omega L_i \ll Z_0$ and $L_i \ll L$, so that

$$j\omega L_i \| Z_0 \approx \frac{\omega^2 L_i^2}{Z_0} + j\omega L_i \quad (20.21)$$

The final equivalent circuit for cavity resonator and diaphragm with source impedance is shown in Fig. 20.7.

The external quality factor follows from equations (20.18) and (20.19)

$$Q_{ext} = \frac{\omega_0 L + \omega_0 L_i}{\omega_0^2 L_i^2 / Z_0} \approx \frac{\omega_0 L}{\omega_0^2 L_i^2 / Z_0} = \frac{b^2}{Y_0^2} \frac{\pi}{2} \frac{\lambda_g^2}{\lambda^2} \tag{20.22}$$

The loaded quality factor is given by

$$\frac{1}{Q_L} = \frac{\omega_0^2 L_i^2 / Z_0 + R}{\omega_0 L + \omega_0 L_i} \approx \frac{\omega_0^2 L_i^2 / Z_0 + R}{\omega_0 L} = \frac{\omega_0^2 L_i^2}{\omega_0 L . Z_0} + \frac{R}{\omega_0 L}$$

$$= \frac{Y_0^2}{b^2} \frac{2}{\pi} \frac{\lambda^2}{\lambda_g^2} + \alpha l \frac{2}{\pi} \frac{\lambda^2}{\lambda_g^2} \tag{20.23}$$

and the unloaded quality factor by

$$Q_0 = \frac{\omega_0 L}{R} \approx \frac{\pi}{2\alpha l} \frac{\lambda_g^2}{\lambda^2} \tag{20.24}$$

Now it can be seen from this equivalent circuit that the resonance frequency is slightly less than ω_0 and it follows that $n\lambda_g$ is slightly greater than $2l$, or l is slightly less than $n\lambda_g/2$. This can also be seen by consideration of the total admittance Y_T in the plane of the diaphragm to the right of the plane of the diaphragm

$$\frac{Y_T}{Y_0} = -j\frac{b}{Y_0} + \frac{1}{\alpha l + j \tan \beta l} = -j\frac{b}{Y_0} + \frac{\alpha l - j \tan \beta l}{(\alpha l)^2 + (\tan \beta l)^2}$$

$$= \frac{\alpha l}{(\alpha l)^2 + (\tan \beta l)^2} - j\frac{b}{Y_0} - j\frac{\tan \beta l}{(\alpha l)^2 + (\tan \beta l)^2} \tag{20.25}$$

At resonance, Y_T/Y_0 should be real; this will clearly be the case if

$$\frac{\tan \beta l}{(\alpha l)^2 + (\tan \beta l)^2} = -\frac{b}{Y_0} < 0 \tag{20.26}$$

i.e. if $\tan \beta l < 0$, or if $l < n\lambda_g/2$.

The input conductance at resonance is obviously

$$\left(\frac{Y_T}{Y_0}\right)_{res} = \frac{\alpha l}{(\alpha l)^2 + (\tan \beta l)^2} \approx \frac{\alpha l}{\tan^2 \beta l} = (\alpha l) \left(\frac{b}{Y_0}\right)^2 \tag{20.27}$$

A numerical example will serve to illustrate the above. A copper waveguide has internal dimensions $22 \cdot 86 \times 10 \cdot 16$ mm and a wall thickness of $0 \cdot 127$ mm. For $\lambda = 32$ mm, we have $\alpha l \approx 4 \times 10^{-4}$, so that for critical coupling $((Y_T/Y_0)_{res} = 1)$ the diaphragm susceptance should be given by $b/Y_0 = 50$. If we choose a circular aperture of diameter d in the centre of A–A′ as coupling hole, then we have

$$\frac{b}{Y_0} \approx \frac{3ab_0 \lambda_g}{2\pi d^3} = 50$$

It follows that $d \approx 4 \cdot 63$ mm.

APPENDICES

APPENDICES

APPENDIX 1
COMPLEX NUMBERS

In this appendix we shall summarize the definitions, rules of calculation and related concepts for complex numbers, which have been widely used in the treatment of wave phenomena, impedances and the like in this book. Complex numbers occur in various algebraic contexts; for example, in quadratic equations when the discriminant of the equation is negative. Consider the equation

$$x^2 + 2x + 2 = 0 \qquad (A1.1)$$

This has the solutions

$$x_{1,2} = -1 \pm \sqrt{-1} \qquad (A1.2)$$

Now $\sqrt{-1}$ is called the imaginary unit and will be denoted by the letter 'i' in this appendix, in line with mathematical usage but contrary to usage in other parts of this book. The solution of the above equation can thus be written

$$x_{1,2} = -1 \pm i \qquad (A1.3)$$

In general, a complex number z may be written

$$z = a + bi \qquad (A1.4)$$

where a and b are normal real numbers and i is again the imaginary units A complex number is thus determined by two real numbers arranged in a definite order. Equation (A1.4) could be regarded simply as a symbolic representation, though we shall see below that the plus sign in the expression really does indicate addition.

Two complex numbers z_1 and z_2 are equal if the two real numbers which define z_1 are equal to the two real numbers which define z_2, order being preserved.

In algebraic language, we write that

$$z_1 = a_1 + b_1 i$$

and

$$z_2 = a_2 + b_2 i$$

are equal only if   (A1.5)

$$a_1 = a_2$$
and $$\qquad b_1 = b_2$$

If the second real number b, defining part of a complex number $z = a+bi$, is zero, then it follows by definition that the complex number z is the same as the real number a. We therefore call a the 'real' part of the complex number $z = a+bi$. If the first real number a defining part of a complex number is equal to zero, so that $z = bi$, the number is said to be imaginary, and bi is called the 'imaginary' part of the complex number. If $b > 0$, bi is said to be positive imaginary, while if $b < 0$, bi is negative imaginary.

The complex number $z = a+bi$ is zero if $a = b = 0$.

The numbers a and b, like any other real number, can assume any real value from $-\infty$ to $+\infty$. This means that all complex numbers can be mapped on a plane called the complex z plane by assigning the complex number $z = a+bi$ to the point in the plane with cartesian co-ordinates (a, b). Conversely, the point in the plane with coordinates (a, b) represents the complex number $(a+bi)$—see Fig. A1.1.

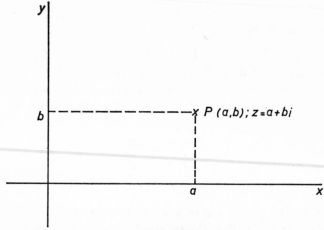

Figure A1.1

We shall now consider the definition of the four basic operations of addition, subtraction, multiplication and division as applied to complex numbers. Our definitions should be such that they give the same results as the normal laws of arithmetic when applied to real numbers.

For addition (subtraction being considered as a special case of addition) we consider the commutative law $\alpha+\beta = \beta+\alpha$, and the associative law, $\alpha+\beta+\gamma = \alpha+(\beta+\gamma)$. Multiplication (division being considered as a special case of multiplication) follows the commutative law $\alpha\beta = \beta\alpha$, the associative law $\alpha\beta\gamma = \alpha(\beta\gamma)$, and the distributive law $\alpha(\beta+\gamma) = \alpha\beta+\alpha\gamma$.

Addition

The sum of two complex numbers $(a+bi)$ and $(c+di)$ is defined as the complex number $\{(a+c)+(b+d)i\}$.

It can easily be verified that this definition satisfies the above-mentioned laws, and it enables us to regard the complex number $(a+bi)$ as the sum of the real number a or $a+0i$ and the imaginary number bi or $0+bi$.

Two complex numbers with the same real part and equal and opposite imaginary parts are said to be 'conjugate' complex numbers. Their sum is real—see Fig. A1.2.

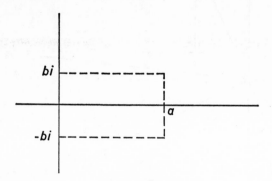

Figure A1.2

If we plot the two numbers $(a+bi)$ and $(c+di)$ and their sum $\{(a+c)+(b+d)i\}$ in the complex plane, one can show that the vector (arrow) from the origin 0 to the point representing the sum can be obtained by combining the vectors from the origin to the points representing the component parts $(a+bi)$ and $(c+di)$, following normal practice in mechanics using the 'parallelogram of forces'—see Fig. A1.3.

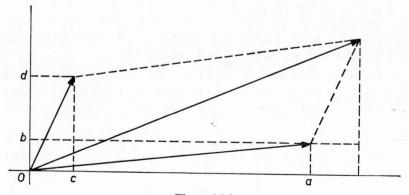

Figure A1.3

The difference of two complex numbers is the number which must be added to the subtrahend to give the minuend, or in algebraic language:

$$(a+bi)-(c+di) = (a-c)+(b-d)i \qquad (A1.6)$$

See also Fig. A1.4.

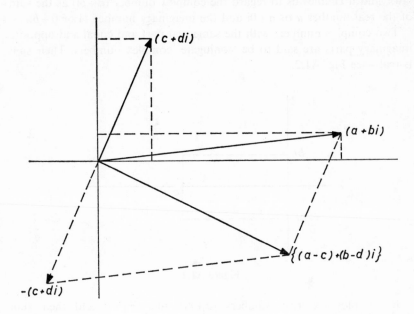

Figure A1.4

Multiplication

The product of two complex numbers $(a+bi)$ and $(c-di)$ is defined as the complex number $\{(ac-bd)+(ad+bc)i\}$. This seems rather a complicated rule, but in fact it is merely the result obtained if the two binomials $(a+bi)$ and $(c+di)$ are multiplied in the normal way

$$(a+bi)(c+di) = ac+adi+bci+bdi^2 = (ac-bd)+(ad+bc)i \qquad (A1.7)$$

making use of the fact that $i^2 = -1$ by definition. It can be verified that this rule does indeed satisfy the commutative, associative and distributive laws.

If one of the complex numbers is zero, the product is also zero. It follows from the definition of complex conjugates that their product is a real number

$$(a+bi)(a-bi) = a^2+b^2 \qquad (A1.8)$$

The square of a complex number can, of course, also be found with the rule for multiplication. This gives

$$(a+bi)^2 = (a^2-b^2)+2abi \qquad \text{(A1.9)}$$

If we use the above definition of multiplication to find the square of i, our result fortunately agrees with the expected. We can continue this

$$(-i)^2 = -1; \; i^3 = -i, \; i^4 = 1, \; i^5 = i, \text{ etc.} \qquad \text{(A1.10)}$$

By the 'quotient' of two complex numbers $(a+bi)$ and $(c+di)$, we mean the complex number by which the divisor $(c+di)$ must be multiplied to get the dividend $(a+bi)$. If we call the quotient $(x+yi)$, we may thus write

$$(x+yi)(c+di) = a+bi$$
or
$$cx-dy+i(dx+cy) = a+bi \qquad \text{(A1.11)}$$

or
$$cx-dy = a$$
$$dx+cy = b \qquad \text{(A1.12)}$$

It follows from equation (A1.12) that

$$x = \frac{ac+bd}{c^2+d^2} \quad \text{and} \quad y = \frac{bc-ad}{c^2+d^2} \qquad \text{(A1.13)}$$

so that

$$\frac{a+bi}{c+di} = \frac{ac+bd}{c^2+d^2} + \frac{bc-ad}{c^2+d^2}i \qquad \text{(A1.14)}$$

In practice, a quotient is determined differently, and the result can be obtained more quickly if we multiply numerator and denominator of the fraction representing the quotient by the complex conjugate of the denominator, thus giving a real denominator

$$\frac{a+bi}{c+di} = \frac{(a+bi)(c-di)}{(c+di)(c-di)} = \frac{ac+bd+(bc-ad)i}{c^2+d^2} \qquad \text{(A1.15)}$$

As an example, we shall consider another special case of multiplication; namely, the determination of the 'square root' of a complex number.

The problem is to find the complex number $(x+yi)$ whose square is equal to the given complex number $(a+bi)$

$$(x+yi)^2 = a+bi$$
or
$$x^2-y^2+2xyi = a+bi \qquad \text{(A1.16)}$$

or
$$x^2-y^2 = a$$
$$2xy = b \qquad \text{(A1.17)}$$

We find without difficulty, by squaring and adding the equations of (A1.17), that

$$(x^2+y^2)^2 = a^2+b^2$$
or
$$(x^2+y^2) = \sqrt{a^2+b^2} \qquad \text{(A1.18)}$$

If we had chosen the negative square root, then x or y would have been imaginary, but according to our definition of a complex number we are only interested in real values of x and y.

With equation (A1.17), (A1.18) gives

$$x^2 = \frac{a}{2} + \tfrac{1}{2}\sqrt{a^2+b^2}$$

$$y^2 = -\frac{a}{2} + \tfrac{1}{2}\sqrt{a^2+b^2}$$

(A1.19)

If we now choose

$$x = +\sqrt{\tfrac{1}{2}a + \tfrac{1}{2}\sqrt{a^2+b^2}}$$

(A1.20)

then according to the second equation of (A1.17), if $b > 0$, we have

$$y = +\sqrt{-\frac{a}{2} + \tfrac{1}{2}\sqrt{a^2+b^2}}$$

(A1.21)

while if $b < 0$

$$y = -\sqrt{-\frac{a}{2} + \tfrac{1}{2}\sqrt{a^2+b^2}}$$

(A1.22)

In this case, we thus find

$$x + yi = \sqrt{\frac{a}{2} + \tfrac{1}{2}\sqrt{a^2+b^2}} \pm i\sqrt{-\frac{a}{2} + \tfrac{1}{2}\sqrt{a^2+b^2}} \quad \text{for } b \gtrless 0 \quad \text{(A1.23)}$$

If we had chosen

$$x = -\sqrt{\frac{a}{2} + \tfrac{1}{2}\sqrt{a^2+b^2}}$$

(A1.24)

we would have found similarly that

$$x + yi = -\sqrt{\frac{a}{2} + \tfrac{1}{2}\sqrt{a^2+b^2}} \mp i\sqrt{-\tfrac{1}{2}a + \tfrac{1}{2}\sqrt{a^2+b^2}} \quad \text{for } b \gtrless 0 \quad \text{(A1.25)}$$

We thus see that there are two complex numbers of opposite sign, equations (A1.23) and (A1.25), whose square is equal to the complex number $(a+bi)$. In calculations with complex numbers, we generally limit ourselves to one of these by selection of the 'principal value' (see below)—in real algebra it is customary to take the positive root of a positive number (not to be confused with the root of an equation)— see Fig. A1.5.

So far, we have written a complex number in the form $z = a+bi$, corresponding to its representation by a point in the complex plane with cartesian coordinates (a, b). However, it can also be defined by polar coordinates, giving another, equally valid notation for complex numbers,

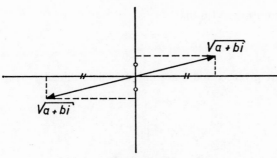

Figure A1.5

which is often more convenient. The radial coordinate of the point $P(a, b)$ is called the modulus of the complex number, and the angle which this radius vector makes with the positive real axis is called the 'argument' of the complex number. The modulus, being the length of a line, is always a positive real number. The modulus of the number z is denoted by $|z|$. A complex number z is uniquely determined by its modulus and argument, but the reverse is not the case, as the argument (angle φ in Fig. A1.6) is determined only for a multiple of 2π. The function argument z (abbreviated arg z) is thus a multi-valued function. In order to obtain a unique notation, we follow the convention that arg z is restricted to the 'principal value', which is defined as follows

$$-\pi < \arg z \leqslant +\pi \tag{A1.26}$$

In line with this convention, we choose one of the two roots determined by the principal value of $(a+bi)$. Now let us write

$$z = x + yi$$
$$|z| = r \tag{A1.27}$$
$$\arg z = \varphi$$

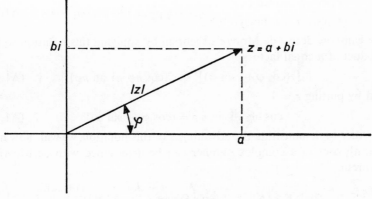

Figure A1.6

It follows from Fig. A1.6 that

$$x = r \cos \varphi$$
$$y = r \sin \varphi \tag{A1.28}$$

If r and φ are given, x and y are uniquely determined

$$r = \sqrt{x^2 + y^2}$$
$$\varphi = \tan^{-1} \frac{y}{x} \tag{A1.29}$$

Note that not every value of φ which satisfied $\tan \varphi = y/x$ satisfies equation (A1.29), so that it can be taken as the argument of $z = x + yi$. For example, if we replace x and y by $-x$ and $-y$, then φ increases by π, but $\tan \varphi$ remains unchanged. However, equation (A1.28) determines φ only for a multiple of 2π, and this ambiguity is resolved by application of equation (A1.26).

We can therefore write the complex number $z = x - yi$ as

$$z = r(\cos \varphi + i \sin \varphi) \tag{A1.30}$$

This notation is not very suitable for the determination of the sums and differences of complex numbers, but it is useful for products and quotients.

Thus $\{r_1(\cos \varphi_1 + i \sin \varphi_1)\}\{r_2(\cos \varphi_2 + i \sin \varphi_2)\}$

$$= r_1 r_2 \{\cos (\varphi_1 + \varphi_2) + i \sin (\varphi_1 + \varphi_2) \tag{A1.31}$$

The modulus of the product of two complex numbers is the product of their moduli, and the argument of the product is the sum of the arguments of the factors.

Similarly, the modulus of the quotient of two complex numbers is the quotient of the moduli of dividend and divisor, and the argument of the quotient is the difference of the arguments of dividend and divisor:

$$\frac{r_1(\cos \varphi_1 + i \sin \varphi_1)}{r_2(\cos \varphi_2 + i \sin \varphi_2)} = \frac{r_1}{r_2} \{\cos (\varphi_1 - \varphi_2) + i \sin (\varphi_1 - \varphi_2)\} \tag{A1.32}$$

We can now derive de Moivre's theorem by applying this notation to the product of n equal factors

$$[r(\cos \varphi + i \sin \varphi)]^n = r^n(\cos n\varphi + i \sin n\varphi) \tag{A1.33}$$

and by putting $r = 1$

$$\cos n\varphi + i \sin n\varphi = (\cos \varphi + i \sin \varphi)^n \tag{A1.34}$$

In order to demonstrate the advantages of this notation, we will show how the nth root of a complex number can be determined with de Moivre's theorem

$$\sqrt[n]{z} = \sqrt[n]{r(\cos \varphi + i \sin \varphi)} = \sqrt[n]{r} \left(\cos \frac{\varphi + 2k\pi}{n} + i \sin \frac{\varphi + 2k\pi}{n} \right) \tag{A1.35}$$

If we give k successively the values $0, 1, 2, \ldots (n-1)$, we find n different complex numbers, all with the same modulus $\sqrt[n]{r}$, whose nth power is equal to $r(\cos \varphi + i \sin \varphi)$. In the complex plane, all these numbers lie on the circle of radius $\sqrt[n]{r}$. The first has the argument φ/n, the next $\varphi/n + 2\pi/n$, the next $\varphi/n + 2(2\pi/n)$, and so on. For example, Fig. A1.7 shows $\sqrt[4]{z}$.

It now follows that

$$\sqrt{i} = \pm \tfrac{1}{2}\sqrt{2}(1+i) \tag{A1.36}$$

However, if we restrict ourselves to the principal value, only one of the roots of equations (A1.35) and (A1.36) remains—namely, that with $k = 0$.

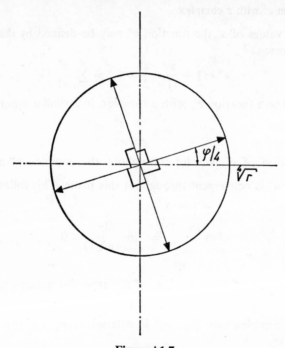

Figure A1.7

Now a few words about series of complex numbers. These are defined in the same way as series of real numbers, and their properties are very similar.

Let S_n be the sum (determined according to the rules of calculation with complex numbers) of the first n terms of the series

$$z_1, z_2, \ldots z_n, \ldots$$

If $S_n \to S$ as $n \to \infty$, the series is said to be convergent, and to have the sum S. In general, of course, both S_n and S will be complex. $\lim_{n \to \infty} S_n = S$ means that for any given small positive number ε, a number N can be found such that for $n > N$ the modulus of the difference between S and S_n is less than ε

$$|S - S_n| < \varepsilon$$

If a series of complex numbers is to be convergent, both the series of the real parts of the terms and that of the imaginery parts must be convergent. Furthermore, a series of complex terms, $\sum z_n$, is convergent if the series of the moduli, $\sum |z_n|$, is convergent.

Other criteria of convergence are the same as for series of real numbers.

The function e^z with z complex

For real values of x, the function e^x may be defined by the following series expansion:

$$e^x = 1 + \frac{x}{1!} + \frac{x^2}{2!} + \ldots = \sum_{n=0}^{\infty} \frac{x^n}{n!} \tag{A1.37}$$

We can define a function e^z, with z complex, in a similar manner

$$e^z = \sum_{n=0}^{\infty} \frac{z^n}{n!} \tag{A1.38}$$

This definition of e^z is valid throughout the complex plane, for the series $\sum_{n=0}^{\infty} z^n/n!$ is convergent throughout this plane. This follows from

$$\lim_{n \to \infty} \frac{\dfrac{|z|^{n+1}}{(n+1)!}}{\dfrac{|z|^n}{n!}} = \lim_{n \to \infty} \frac{|z|}{(n+1)} = 0 \tag{A1.39}$$

In the same way as for real values of the exponent we may write

$$e^{z_1} e^{z_2} = e^{z_1 + z_2} \tag{A1.40}$$

where the complex sum $(z_1 + z_2)$ is determined by the usual complex number rules.

Let us consider the function $e^{i\alpha}$ with α real. If we expand this according to equation (A1.38) and separate the real and imaginary terms, we find

$$\begin{aligned}
e^{i\alpha} &= 1 + i\alpha - \alpha^2/2! - i\alpha^3/3! + \alpha^4/4! + i\alpha^5/5! - \alpha^6/6! + \ldots \\
&= (1 - \alpha^2/2! + \alpha^4/4! - \alpha^6/6! + \ldots) + i(\alpha - \alpha^3/3! + \alpha^5/5! - \ldots) \\
&= \cos \alpha + i \sin \alpha
\end{aligned} \tag{A1.41}$$

This is the proof of Euler's relation for real α

$$e^{i\alpha} = \cos \alpha + i \sin \alpha \tag{A1.42}$$

It follows from the definition of the modulus and from equation (A1.42) that

$$|e^{i\alpha}| = \sqrt{\cos^2 \alpha + \sin^2 \alpha} = 1$$

The argument of $e^{i\alpha}$ is α, so that $e^{i\alpha}$ represents a point on the unit circle in the complex plane. For example

$$e^{\pi i/2} = i \qquad e^{\pi i} = -1 \qquad e^{3\pi i/2} = -i \qquad e^{2\pi i} = e^{4\pi i} = 1$$

(see also Fig. A1.8).

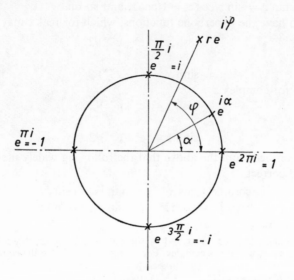

Fig. A1.8.

Now we have seen above that a complex number can be written not only as $z = x + yi$ but also in terms of the modulus and argument, $z = r(\cos\varphi + i\sin\varphi)$. It follows from equation (A1.42) that the latter notation is identical to

$$z = r\,e^{i\varphi} \tag{A1.43}$$

This is, in fact, the most usual way of writing a complex number in terms of its modulus and argument.

We may conclude that e^z is a periodic function with the imaginary period $2\pi i$; in other words, e^z does not change when z changes by $2\pi i$.

We know that for real x

$$e^{ix} = \cos x + i\sin x$$

and hence

$$e^{-ix} = \cos x - i\sin x$$

so that

$$\cos x = (e^{ix} + e^{-ix})/2 \tag{A1.44}$$

and

$$\sin x = (e^{ix} - e^{-ix})/2i \tag{A1.45}$$

We can define the functions cos z and sin z for z complex in an analogous manner

$$\cos z = (e^{iz} + e^{-iz})/2 \tag{A1.46}$$

$$\sin z = (e^{iz} - e^{-iz})/2i \tag{A1.47}$$

and further, by analogy with the normal goniometric formulae for real variables, $\tan z = \sin z/\cos z = 1/\cot z$, and so on.

We also have the hyperbolic functions, which for real x may be written

$$\cosh x = (e^x + e^{-x})/2 \tag{A1.48}$$

and

$$\sinh x = (e^x - e^{-x})/2 \tag{A1.49}$$

Similarly, for complex z

$$\cosh z = (e^z + e^{-z})/2 \tag{A1.50}$$

and

$$\sinh z = (e^z - e^{-z})/2 \tag{A1.51}$$

It can be verified from the above that the following widely used relations are indeed correct

$$\begin{array}{ll} \cosh ix = \cos x & \sinh ix = i \sin x \\ \cos ix = \cosh x & \sin ix = i \sinh x \end{array} \tag{A1.52}$$

where x is real.

Note: We know that for real x, $-1 \leqslant \sin x \leqslant +1$. However, this limitation does not apply to sin z, where z is complex. For example, let us solve the equation

$$\sin z = 2$$

By definition (equation A1.47)

$$\sin z = \frac{e^{iz} - e^{iz}}{2i}$$

now let us put $e^{iz} = t$. We then have

$$\frac{t - 1/t}{2i} = 2 \quad \text{or} \quad t^4 - 4it - 1 = 0$$

with solutions

$$t_{1,2} = 2i \pm \sqrt{-3} = i(2 \pm \sqrt{3})$$

This equation has the two solutions

$$t_1 = e^{iz_1} = i(2 + \sqrt{3})$$

and

$$t_2 = e^{iz_2} = i(2 - \sqrt{3})$$

The solutions of the original equation (sin $z = 2$) may thus be written

$$z_1 = \frac{\pi}{2} + 2k\pi - i \ln (2 + \sqrt{3})$$

and

$$z_2 = \frac{\pi}{2} + 2k\pi + i \ln (2 + \sqrt{3})$$

where k is zero or an integer.

The second root may be obtained by subtracting the first from π, as with the roots of $\sin x = a$ with $|a| \leqslant 1$.

The last operation on complex numbers we shall describe is the taking of logarithms of complex numbers. We have already used this above in solving $e^{iz} = i(2 + \sqrt{3})$.

What we are looking for is the natural logarithm of $z = r\,e^{i\varphi}$, generally written $\ln z$. This is defined, by analogy with the definition for real numbers, as the power to which 'e' must be raised to give z. Let us call this power w, which will in general be complex. We may then write

$$e^w = z = r\,e^{i\varphi} = e^{\ln r}\,e^{i\varphi} = e^{\ln r + i\varphi} \qquad (A1.53)$$

or

$$w = \ln z = \ln\{r\,e^{i\varphi}\} = \ln r + i\varphi \qquad (A1.54)$$

This may be stated in words as follows: the natural logarithm of a complex number is equal to the natural logarithm of the modulus r of that complex number, plus 'i' times the argument φ.

Since the argument φ is limited to the principal value, this defines $\ln z$ uniquely. For example

$$\ln i = \ln|i| + i\pi/2 = i\pi/2$$
$$\ln(-i) = \ln|-1| - i\pi/2 = -i\pi/2$$
$$\ln(-1) = \ln|-1| + i\pi = i\pi$$

Logarithms of complex numbers share the properties of real positive numbers, e.g.

$$\ln z_1 z_2 = \ln z_1 + \ln z_2, \text{ etc.}$$

APPENDIX 2
THE SMITH CHART

In our discussion of parallel-wire transmission lines, such as coaxial, Lecher, strip, etc., and rectangular waveguides we have seen that certain concepts such as reflection coefficient, standing-wave ratio, input impedance of a length of line with terminal impedance Z_l, etc., are of general application to all types of transmission line. In empty parallel-wire lines, the wavelength is the same as in free space, while in empty rectangular waveguides it is not. In parallel-wire transmission lines, the voltage and the current vary along the line, while in rectangular waveguides it is the electric and magnetic fields which vary. In both cases, these variations can be described by assigning a characteristic impedance, an attenuation constant and a phase constant to the line. A knowledge of the way in which these parameters vary along the line is invaluable for the design of transmission-line components and for an understanding of the behaviour of lines in general.

Apart from the equations derived in the foregoing chapters, graphic aids often give an insight into the situation more simply and flexibly than is possible with algebraic manipulations, which are often long and complicated. One of these graphic aids is the Smith chart or diagram, based on plotting the reflection coefficient $r = |r| \, e^{j\varphi}$ in polar coordinates in the complex plane (Fig. A2.1). It follows from the definition of the reflection coefficient that $|r| \leqslant 1$, so that we need consider only that part of the complex plane lying within the unit circle.

We found the following expression for the reflection coefficient at $z = l$ of a length of line of characteristic impedance Z_0 terminated by an impedance of Z_l at $z = l$

$$r(0) = \frac{Z_l - Z_0}{Z_l + Z_0} = \frac{Z_l/Z_0 - 1}{Z_l/Z_0 + 1} = |r| \, e^{j\varphi} \qquad (A2.1)$$

A short distance along the line from this termination, the modulus of the reflection coefficient remains the same (as long as no discontinuities occur

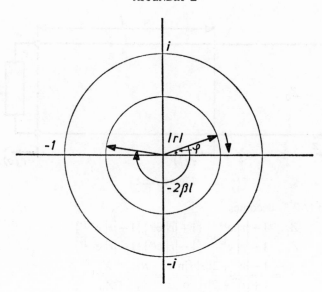

Figure A2.1

in the intervening line). Only the phase changes, according to the equation

$$r(l) = r(0)\, e^{-2j\beta l} \tag{A2.2}$$

(See also Fig. A2.2.)

In the complex plane, this corresponds to a clockwise rotation of the vector r through an angle of $2\beta l$ radians from its original position.

We shall now express the input impedance in terms of the reflection coefficient at the point where the input impedance is determined, and as before find

$$
\frac{Z_i}{Z_0} = \frac{Z_l + jZ_0 \tan \beta l}{Z_0 + jZ_l \tan \beta l} = \frac{Z_l + jZ_0 \dfrac{e^{j\beta l} - e^{-j\beta l}}{2j} \cdot \dfrac{2}{e^{j\beta l} + e^{-j\beta l}}}{Z_0 + jZ_l \dfrac{e^{j\beta l} - e^{-j\beta l}}{2j} \cdot \dfrac{2}{e^{j\beta l} + e^{-j\beta l}}}
$$

$$
= \frac{Z_l(e^{j\beta l} + e^{-j\beta l}) + Z_0(e^{j\beta l} - e^{-j\beta l})}{Z_0(e^{j\beta l} + e^{-j\beta l}) + Z_l(e^{j\beta l} - e^{-j\beta l})} = \frac{(Z_l + Z_0)\, e^{j\beta l} + (Z_l - Z_0)\, e^{-j\beta l}}{(Z_l + Z_0)\, e^{j\beta l} - (Z_l - Z_0)\, e^{-j\beta l}}
$$

$$
= \frac{1 + \dfrac{Z_l + Z_0}{Z_l + Z_0}\, e^{-2j\beta l}}{1 - \dfrac{Z_l - Z_0}{Z_l + Z_0}\, e^{-2j\beta l}} = \frac{1 + r(0)\, e^{-2j\beta l}}{1 - r(0)\, e^{-2j\beta l}} = \frac{1 + r(l)}{1 - r(l)} \tag{A2.3}
$$

where

$$
r(l) = r(0)\, e^{-2j\beta l} = |r|\, e^{j\varphi}\, e^{-2j\beta l} = |r|\, e^{j(\varphi - 2\beta l)}
$$
$$
= |r|\, e^{j\alpha} \tag{A2.4}
$$

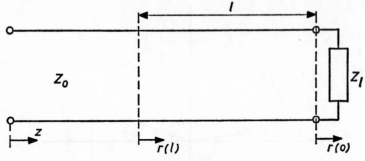

Figure A2.2

Equation (A2.3) becomes

$$\frac{Z_i}{Z_0} = \frac{1+|r|\,e^{j\alpha}}{1-|r|\,e^{j\alpha}} = \frac{\{1+|r|\,e^{j\alpha}\}\{1-|r|\,e^{-j\alpha}\}}{\{1-|r|\,e^{j\alpha}\}\{1-|r|\,e^{-j\alpha}\}}$$

$$= \frac{1-|r|^2+2j|r|\sin\alpha}{1+|r|^2-2|r|\cos\alpha} = \frac{R_i}{Z_0}+j\frac{X_i}{Z_0} \qquad (A2.5)$$

The real and imaginary parts of the normalized input impedance can thus be written

$$\frac{R_i}{Z_0} = \frac{1-|r|^2}{1+|r|^2-2|r|\cos\alpha} \qquad (A2.6)$$

$$\frac{X_i}{Z_0} = \frac{2|r|\sin\alpha}{1+|r|^2-2|r|\cos\alpha} \qquad (A2.7)$$

If the terminal impedance is known, or if $|r|$ and φ are known (which comes to the same thing in equation A2.1), then the reflection coefficient is determined for any point along the line, and the input impedance is also determined according to equations (A2.6) and (A2.7). It is thus always possible to calculate R_i/Z_0 and X_i/Z_0 from these equations, but the algebraic manipulations involved are complicated, and it would be convenient if these could be replaced by a rapid graphical method. What we want in fact is a graph of the reflection coefficient in the complex plane, giving lines of constant R_i/Z_0 and X_i/Z_0 for various values of the reflection coefficient. To obtain this, we first write

$$\frac{R_i}{Z_0} = \text{constant} = p \qquad\qquad \frac{X_i}{Z_0} = \text{constant} = q$$

$$p = \frac{1-|r|^2}{1+|r|^2-2|r|\cos\alpha} \qquad q = \frac{2|r|\sin\alpha}{1+|r|^2-2|r|\cos\alpha}$$

Changing to cartesian coordinates

$$\begin{cases} x = |r|\cos\alpha \\ y = |r|\sin\alpha \end{cases}$$

we find

$$p = \frac{1-x^2-y^2}{1+x^2+y^2-2x} \qquad q = \frac{2y}{1+x^2+y^2-2x}$$

$$x^2(1+p)-2px+y^2(1+p) = (1-p) \qquad x^2-2x+1+y^2 - \frac{2}{q}y = 0 \quad \text{(A2.8)}$$

$$\left(x - \frac{p}{1+p}\right)^2 + y^2 = \frac{1}{(1+p)^2} \qquad (x-1)^2 + \left(y - \frac{1}{q}\right)^2 = \frac{1}{q^2}$$

Now the equation $(x-a)^2+(y-b)^2 = c^2$ represents a circle in the xy plane with centre at (a, b) and radius c.

It follows that the lines of constant R_i/Z_0 form a family of circles with centre $[p/(1+p), 0]$ (on the real axis) and radius $|1/(1+p)|$. All these circles pass through the point $(1, 0)$, and as the value of p varies from 0 to ∞, the centre of the circle moves from $(0, 0)$ to $(1, 0)$.

The whole region in which R_i/Z_0 is constant lies within the unit circle (Fig. A2.3).

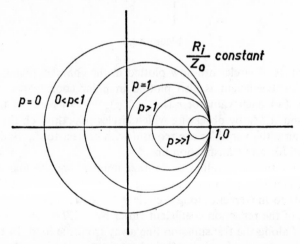

Figure A2.3

The lines of constant X_i/Z_0 form a family of circles with centre $(1, 1/q)$ on a vertical line through $(1, 0)$, and radius $|1/q|$. Here too, each circle of the family passes through the point $(1, 0)$. The parameter q can vary from $-\infty$ to $+\infty$. For $q = 0$, the circle degenerates into the real axis, while for $q \to \pm\infty$ the circle degenerates into the point $(1, 0)$. For $q = \pm1$, the circles pass through $(0, \pm i)$, $q > 0$ gives circles above the real axis, and $q < 0$ circles below the real axis (Fig. A2.4).

Note: When $X_i = 0$ and $R_i = Z_0$, there is a travelling wave in the transmission line. In the Smith chart (Figs. A2.3 and A2.4), this corresponds to the centre of the unit circle ($|r|=0$).

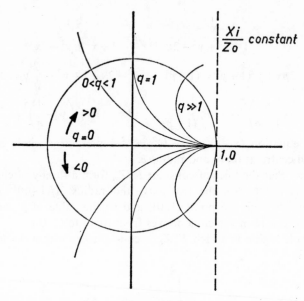

Figure A2.4

The families of circles are now plotted in the complex plane, in which the reflection coefficient is also plotted in polar coordinates inside the unit circle. For each value of r ($|r|$ and φ), we can read off the values of R_i/Z_0 and X_i/Z_0 by observing which circles pass through the required point. Because the variables are reduced with respect to Z_0, Smith charts can be used for any transmission line.

If we want to know values at another point along the line, we make use of the fact that $|r|$ remains constant, so that passage along the line corresponds to movement along a circle with centre $(0, 0)$ and radius $|r|$. The phase of the reflection coefficient varies by $-2\beta l$ radians as we move a distance l along the transmission line away from the load. In the Smith chart, this corresponds to a clockwise rotation through $2\beta l$ radians. Having arrived at the required point, we read off the new values of R_i/Z_0 and X_i/Z_0. For the sake of convenience, the Smith chart is provided with a scale round the unit circle which gives this angle of rotation expressed in wavelengths

$$2\beta l = 2 \frac{2\pi}{\lambda} l = 4\pi \frac{l}{\lambda}$$

If l is half a wavelength, the angle of rotation is 2π radians, i.e. we return to our starting point. This is what happens in practice: in a transmission line with a standing wave, conditions repeat themselves at intervals of a half wavelength (see Fig. A2.5). The circumference of the unit circle in the Smith chart is therefore graduated in fractions of half a wavelength. Clockwise rotation corresponds to movement along the waveguide away from the load (towards the source), while anticlockwise rotation corresponds to movement towards the load (away from the source).

In fact, Smith charts do not give the reflection coefficient as such, but rather the standing-wave ratio, which is equal to R_i/Z_0. The scale in the diagram is, of course, non-linear since it is based on a linear scale for the reflection coefficient and $\eta = (1+|r|)/(1-|r|)$.

We still need to determine where the diagram should start for reflection coefficients with zero phase. It would be logical to take the line along which the reflection coefficient is real, i.e. from (0, 0) to (1, 0), as representing zero phase, and to start counting rotation in wavelengths along the circumference of the unit circle from there. However, the convention is that we start at the opposite point $(-1, 0)$, because this point represents a voltage minimum along the transmission line and it is easier to determine the minimum in the standing wave, which is often quite sharp, than the maximum, where the curvature is much less. An example of the Smith chart is shown in Fig. A2.6 (p. 274).

Before finishing this Appendix with a sample calculation, we should discuss the representation of admittances in this diagram, which is based on impedances. If a series impedance is added somewhere along the transmission line, all we need do to find the overall impedance is to add the extra impedance to the input impedance at the spot in question. It would be convenient if the addition of an impedance in parallel with the line could be dealt with in terms of admittances, so that here too the result could be found by means of a simple addition.

We shall now examine how the equivalent admittance along the line can be found from a given impedance by using the Smith chart. In general, the reduced admittance is given by equation (A2.5).

$$\frac{Y_i}{Y_0} = \frac{G_i+jB_i}{Y_0} = \frac{1}{Z\,l/Z_0} = \frac{1-|r|\,e^{j\alpha}}{1+|r|\,e^{j\alpha}}$$

$$= \frac{1-|r|^2-2j|r|\sin\alpha}{1+|r|^2+2|r|\cos\alpha} \tag{A2.9}$$

So that

$$\frac{G_i}{Y_0} = \frac{1-|r|^2}{1+|r|^2+2|r|\cos\alpha} \qquad \frac{B_i}{Y_0} = \frac{-2|r|\sin\alpha}{1+|r|^2+2|r|\cos\alpha} \tag{A2.10}$$

Changing to cartesian coordinates

$$\begin{cases} x = |r|\cos\alpha \\ y = |r|\sin\alpha \end{cases}$$

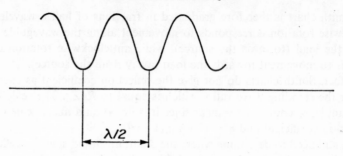

Figure A2.5

we find

$$\frac{G_i}{Y_0} = \frac{1-x^2-y^2}{1+x^2+y^2+2x} \qquad \frac{B_i}{Y_0} = \frac{-2y}{1+x^2+y^2+2x} \qquad \text{(A2.11)}$$

If we denote the lines $G_i/Y_0 = $ constant by p', and those for $B_i/Y_0 = $ constant by q', we may write

$$1-x^2-y^2 = p'+p'x^2+p'y^2+2p'x \qquad -\frac{2}{q}y = 1+x^2+y^2+2x$$

or

$$\left(x+\frac{p'}{1+p'}\right)^2 + y^2 = \left(\frac{1}{1+p'}\right)^2 \qquad (x+1)^2 + \left(y+\frac{1}{q'}\right)^2 = \left(\frac{1}{q'}\right)^2 \qquad \text{(A2.12)}$$

The lines $G_i/Y_0 = $ constant $= p'$ form a family of circles with centre $(-p'/(1+p'), 0)$ and radius $|1/(1+p')|$, while the lines $B_i/Y_0 = $ constant $= q'$ form a family of circles with centre $(-1, -1/q')$ and radius $|1/q'|$. If we compare these families of circles with those representing the real and imaginary parts of the impedance, we find that the admittance circles can be obtained from the impedance circles by replacing x and y by $-x$ and $-y$, i.e. by rotation of π radians about the centre of the Smith chart; we must also, of course, replace p by p' and q by q'. The values of $R_i/Z_0 = p$ and $X_i/Z_0 = q$ given round the impedance circles should thus be read as values of $G_i/Y_0 = p' = p$ and $B_i/Y_0 = q' = q$ in the present case.

The above rotation round the centre of the diagram is equivalent to rotation through a quarter wavelength. In other words, a given impedance at the point $z = l$ in a transmission line is transformed to its inverse at the point $z = l \pm \frac{1}{4}\lambda_g$ (a series resonance at a given spot behaves as a parallel resonance a quarter wavelength further).

That part of the chart which gave a negative reactance for impedances will give positive susceptances when the chart is used for admittances, and vice versa.

Frequency is not given in the chart. If measurements are made as a function of frequency, the latter should be written in the chart alongside the curve plotted.

As we rotate the chart (or move along the transmission line), we see that the impedance assumes a real value twice in one complete revolution, or half-wavelength shift along the waveguide (one high value and one low). If desired, the guide can be matched by transforming the high or the low value to the characteristic impedance by means of a quarter-wave transformer at one of these points.

In addition, the real part of the impedance will be equal to the characteristic impedance ($R_i/Z_0 = 1$) at two points per full rotation. At these points, matching can be obtained by introducing an extra positive or negative reactance into the waveguide.

A specimen calculation will now complete this section. The curve in the Smith chart is shown in Fig. A2.6, and the corresponding circuit diagram is given in Fig. A2.7. A transmission line of characteristic impedance $Z_0 = 50 \ \Omega$ is terminated by an impedance $Z_l = 100 - j175 \ \Omega$, so that $Z_l/Z_0 = 2 \cdot 0 - j3 \cdot 5$ (point A in the chart). Let the measuring frequency be $f = 1000$ MHz. For the sake of convenience, we shall let the wavelength in the line be equal to that in free space (coaxial line). At a distance of 9 cm ($= 0 \cdot 3 \ \lambda$) from the terminal load, an extra series impedance of $+j26 \cdot 5$ is introduced; reduced with respect to Z_0, this gives $+j0 \cdot 53$. The input impedance at this spot, just to the right (Fig. A2.7) of this extra reactance, is found by moving a distance of $0 \cdot 3 \ \lambda$ from A towards the generator along an arc of a circle with the origin O of the chart as centre and OA as radius. This gives point B. We started at $0 \cdot 284 \ \lambda$ and finish at $(0 \cdot 284 + 0 \cdot 3) \ \lambda = 0 \cdot 584 \ \lambda$, or $0 \cdot 084 \ \lambda$. It will be seen that the input impedance at B is $(0 \cdot 15 + j0 \cdot 57)$. Together with the extra reactance of $j0 \cdot 53$, we obtain $(0 \cdot 15 + j1 \cdot 1)$, represented by point C. Now the problem is to match this impedance to the characteristic impedance of the line. This means that we must finish at the origin O of the chart. As we then move along the line (rotate in the chart), the input impedance would remain constant, with $Z_i/Z_0 = 1$.

The matching can be carried out in several different ways:

1. From C, we rotate in the direction of the source until the real part of the input impedance becomes equal to Z_0 (point D). It will be seen that this point is a distance $(0 \cdot 209 - 0 \cdot 133) \ \lambda = 0 \cdot 076 \ \lambda$ to the left of the added series impedance (Fig. A2.7), i.e. $9 + 2 \cdot 28 = 11 \cdot 28$ cm from the terminal impedance. At this point, the input impedance is $(1 + j3 \cdot 5)$, so that we must add a series impedance $-j3 \cdot 5$ here, i.e. a series reactance of $-j3 \cdot 5 \times 50 = -j175$ in absolute value.

2. We rotate from C in the direction of the source until the input impedance becomes real (in principle, there are two possible points to

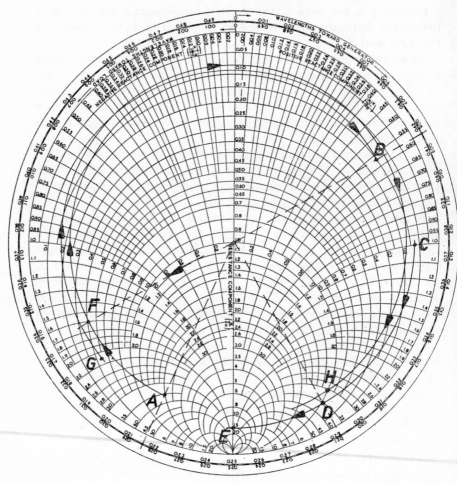

Figure A2.6

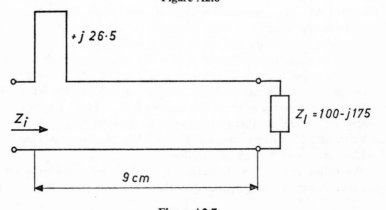

Figure A2.7

choose from, but we use the first). This brings us to point E (0·041 λ further than D), where $R_i/Z_0 = 14$. Matching can be done by installing a quarter-wave transformer with characteristic impedance $Z_0' = \sqrt{14}\,Z_0$ at this point.

If, instead of the series impedance, we had placed a reduced reactance of $+2j$ in parallel with the transmission line at point B (9 cm from the terminal load), we could also calculate the matching conditions from the chart. The presence of this parallel reactance of $2j$ means that a susceptance of $-0·5j$ must be added to the admittance at this point. This admittance is found by rotating point B π radians around O, giving point F where the admittance is $(0·45-j1·6)$. Adding $-0·5j$, we arrive at point G $(0·45-j2·1)$. In order to match, we rotate towards the source until the real part of the admittance is equal to Y_0 (point H), and then add the necessary susceptance $(-j3·0)$ in parallel with the line:

APPENDIX 3 SOLID-STATE MICROWAVE COMPONENTS

In this book, we have not considered the generation, amplification and radiation of microwaves, but have concentrated on transmission line techniques without going into technical details of modern designs. It will, however, be useful to devote a few words to some of the solid-state components used with microwaves. This is not the place for an exhaustive discussion; for further details, the reader is referred to the literature.

Ferrite components

One group of components is formed by lengths of transmission line partially filled with ferrite as magnetic material. A number of these components have been in general use for many years.

Isolators

Isolators consist of lengths of transmission line (such as rectangular waveguides, coaxial lines, strip lines) containing a piece of ferrite of a known composition and dimension at a fixed point, sometimes accompanied by a piece of dielectric material. The field configuration is consequently modified at this point, with the result that the propagation constants in the two directions of the line are different; the isolator is thus a non-reciprocal element. The component transmits electromagnetic waves in one direction with little or no attenuation, while in the opposite direction the waves are strongly attenuated.

Components of this type are used to prevent changes which may occur in parts of the system behind the isolator from spreading to parts in front of the isolator. For example, they are often used to decouple the source or the load from the rest of the system.

Their operation is frequency-dependent, but wide-band versions can be made.

Circulators

The circulator, which is also a non-reciprocal element containing ferrite materials, has three or more ports ('terminal pairs'). The power applied to one of these ports is 'passed' to the other ports in an order determined by an externally applied magnetic field. This 'passage' of power should be seen as follows: if one port is externally short-circuited, or terminated with a reactive load, the energy coming from the previous port is totally reflected and offered to the next port in a predetermined direction of circulation. If, on the other hand, the port is provided with a matched load, it absorbs all the power coming from the previous port so that any subsequent ports, and in particular the original input, receive nothing. The reverse attenuation, in the direction opposed to the imposed direction of circulation, is high (of the order of tens of decibels), while the forward attenuation from one arm to the next is low (of the order of tenths of a decibel).

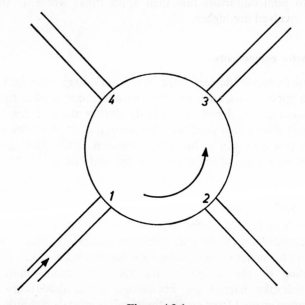

Figure A3.1

Circulators, like isolators, are used to decouple various parts of transmission line systems. They are particularly suitable for use with negative-resistance amplifiers. The source (aerial) to be amplified is then connected to port 1, the negative-resistance amplifier (which is, in principle, a one-port circuit) to port 2 and the receiver to port 3 (Fig. A3.1). A fourth port may be included, with a matched load, to provide extra decoupling between the aerial and the receiver if either should not be completely reflection-free. Energy from port 1 is offered to port 2, where it is reflected with amplification by the negative-resistance amplifier (reflection coefficient < -1).

Like the isolator, the circulator is frequency-dependent, but wide-band versions are available.

Control components

A number of other ferrite components, some of which are also non-reciprocal, can be grouped under the name of control components. Much the same basic design is used for all such components, regardless of whether their function is attenuation, phase shifting, limiting, modulating or filtering. They are used, for example, in duplexer units for radar (with or without limiters), phase shifters for array antennas, power switches and the like. Many of these functions can also be fulfilled by semiconductor elements, but ferrite components tend to be used for high-power applications whilst semiconductors find their applications where the frequency variations involved are higher.

Semiconductor components

While semiconductor devices have for a very long time had various microwave applications, such as detectors and mixer diodes, their range of application has widened increasingly during the last few decades. This is largely due to their small size, low weight, great reliability and other advantages (for example, as low-noise amplifiers). We shall now briefly discuss various groups of semiconductor applications.

Receivers

Superheterodyne receivers with mixer stages: Point-contact diodes, which have long been used for detection and mixing purposes, are still undergoing continual refinement, particularly regarding the noise figure attainable for the mixer stage (using germanium instead of silicon and better understanding of the design of the rectifying metal–semiconductor contact). Schottky barrier (or hot-carrier) mixer diodes are gaining more and more ground from conventional mixer crystals. Schottky diodes, which are made by planar techniques, have better noise properties (5–7 dB, depending on frequency) and are less likely to break down under the influence of load peaks. They are also very suitable for use in mixer stages designed with hybrid integrated circuits for very small strip-line systems.

Microwave pre-amplifiers: Another method of improving the noise properties of the receiver involves the use of low-noise pre-amplifiers. We shall mention a number of types here.

Tunnel-diode amplifier: A tunnel-diode amplifier is a negative-resistance amplifier which is nearly always used in conjunction with a circulator, and gives reasonably low noise figures (3–6 dB, depending on frequency and diode material) over fairly wide frequency ranges. However, the noise characteristics are not much better than those of modern mixer stages. Moreover, applications are limited by a rather low saturation level. The negative resistance is due to the fact that the I–V characteristic of the tunnel diode has a negative slope over part of its length (Fig. A3.2).

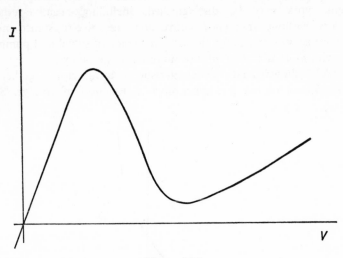

Figure A3.2

As a result of 'tunnelling' of electrons through the potential barrier, a 'tunnel current' is produced which increases to a maximum as the forward d.c. voltage across the crystal is increased, and then falls off to zero as the forward voltage increases further (owing to a lack of free levels on the far side of the potential barrier). The second rising part of the I–V characteristic is due to the normal forward diffusion current, which attains a measurable positive value before the tunnel current has fallen to zero, so that a finite 'valley current' is produced. With a suitable d.c. bias, the tunnel diode acts as a negative resistance. When it is included in a microwave resonance circuit, amplification can be achieved, a typical power gain being about 15 dB.

Parametric amplifiers: Parametric amplifiers can be used where very low noise levels are required. Certain versions, which still often require cooling, give noise characteristics comparable to those of a maser: effective input noise temperatures of a few tens or less Kelvin units or noise figures of a few tenths of a decibel. Like the tunnel-diode amplifier, the parametric amplifier is in general a negative-resistance type, and circulators should preferably also be used. The negative resistance is due to a mixing process, via a reactance which varies periodically with time (in practice, a non-linear capacitance formed by a semiconductor diode, the varactor, biased in the reverse direction), involving the signal to be amplified and a 'pump' signal. This pump signal generally has a much higher frequency than the signal to be amplified. The difference (or sum) frequency, known as the idler frequency, feeds a resonance circuit which should be 'seen' by the varactor. The circuits are fairly complex. Many different types may be distinguished, including negative-resistance amplifiers without frequency conversion, negative-resistance mixers, positive-resistance mixers for the sum frequency of signal and pump, and others with more than one idler or more than one pump.

Fig. A3.3 shows the I–V characteristic of a varactor, together with the capacitance of the p–n junction as a function of voltage. Super-

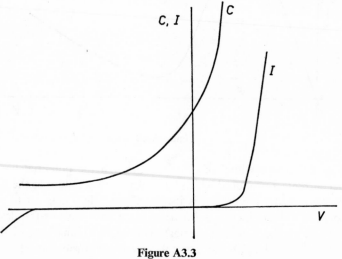

Figure A3.3

position of the pump signal on a negative bias voltage gives the required periodically varying reactance; this principle can be used down to millimetre wavelengths. In view of developments in gallium arsenide planar varactors, very low noise figures will be achieved with amplifiers that require little cooling.

Transistor amplifiers: It is possible to manufacture bipolar transistors that can be used for the lower microwave frequencies (up to about 4–5 GHz). Both silicon and germanium are used as the semiconductor material. The noise figure attained can vary from about 4 dB at 2–3 GHz to about 5 dB at 4 GHz, while the power amplification amounts to about 10 to 4 dB per stage, depending on the frequency. Both the performance at these frequencies and the frequency range itself are constantly being improved so that we may expect the transistor to become much more widely used in all kinds of microwave circuits.

Characteristics of the various types of amplifier discussed above are summarized in Fig. A3.4.

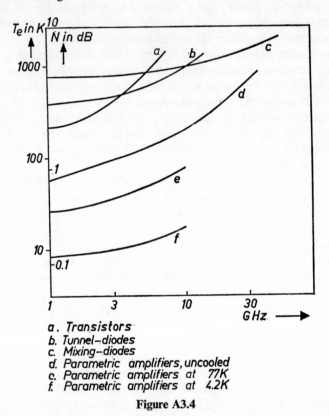

a. Transistors
b. Tunnel-diodes
c. Mixing-diodes
d. Parametric amplifiers, uncooled
e. Parametric amplifiers at 77K
f. Parametric amplifiers at 4.2K

Figure A3.4

Other solid-state amplifiers: Making use of the Gunn effect and avalanche processes (see below), research is being carried out into the design of negative-resistance power amplifiers. At present the noise figures obtained are, however, rather high.

Energy sources

Varactor multiplier chains: The non-linear reactance provided by a varactor with negative bias can be used for the multiplication of the frequency of a signal from a transistor oscillator, for example, at VHF or UHF frequencies. In general, a number of doubler, tripler or even quadrupler stages will be arranged in cascade to give the desired output frequency. Such a chain can be given a very high frequency stability by synchronization with a crystal oscillator. Silicon varactors give the best results down to wavelengths of about 30 mm (fairly high breakdown voltages); at lower wavelengths, gallium arsenide varactors will be preferable in view of their higher cut-off frequency, which will allow the output power to attain higher values. The total number of stages is sometimes quite large, so that with the filters and the extra resonance circuits at idler frequencies the circuit is quite complex.

Step-recovery diodes (also called Boff diodes, snap-off diodes, punch-through varactors, charge-storage diodes and piecewise linear diodes) are more suitable for higher single-stage frequency multiplication (by a factor of 10 or more). The output power of these circuits is often lower than that of a normal varactor chain, and higher demands are made on the filters, but the circuits are much simpler. These circuits find their main application for local oscillators with output powers of the order of tens of milliwatts.

Varactor chains can give output powers of the order of 10 W at 1 GHz, and 1 W at 10 GHz. This power can be increased by the use of 'stacked' varactors, or with a number of diodes in parallel on the same substrate and in the same encapsulation.

Transistor oscillators: We have already mentioned the role of the transistor oscillator as the input stage of a varactor frequency-multiplier chain (giving tens of watts at UHF frequencies). The transistor can, of course, also be used directly as a microwave oscillator. At present, its use in this field is limited to the lower microwave frequencies; output powers of a few tens of watt at 1 GHz are possible with multi-emitter silicon transistors decreasing to about 1 watt at about 4 GHz. Higher output frequencies can be reached by using the non-linear collector capacitance of the transistor as a varactor in an internal frequency-multiplication process. Preliminary results have been reported in the literature about transistors oscillating in their fundamental mode up to frequencies in the region of 10 GHz.

Gunn-effect oscillators: If a sufficiently high d.c. voltage is applied to n-type gallium arsenide, microwave oscillations can be produced. These volume instabilities (negative-resistance effects) are due to the special structure of the conduction bands in gallium arsenide (subsidiary energy minima associated with low mobilities, alongside the main minimum with high mobility). The form of the cavity resonator in which the biased semiconductor is placed has a great effect on the operating frequency, the tuneability and the output power.

The Gunn effect can be used in various ways in the design of microwave oscillators. In the Gunn effect proper, the transit time of the charge carriers through the semiconductor crystal plays an important role in determining the final operating frequency of the oscillator. However, in another form of the Gunn effect (limited space-charge accumulation, or LSA mode), this transit time plays a minimal role. Larger semi-conductor crystals can then be used to increase the output power. This is done particularly at millimetre wavelengths.

In general, oscillators built on these principles can now give continuous output powers for single devices of the order of 1 watt round about 10 GHz. Output powers of hundreds of watt are possible in pulse operation round about 10 GHz and of kilowatt at about 1 GHz, and combinations of a number of devices can be used to increase these levels.

Avalanche-diode oscillators: The avalanche oscillator is another micro-wave source based on a negative-resistance effect in semiconductors, and using silicon, germanium and gallium arsenide. In its original form, proposed by Read, this oscillator made use of a p–n–i–n structure (the Read diode), but a later form which is more widely used makes use of p–n junctions in silicon, which are easier to manufacture. These oscillators are sometimes called Impatt (Impact avalanche and transit time) oscillators.

A breakdown mechanism gives rise to a charge wave in a narrow region (highly homogeneous structures are essential) which has a phase lag of about 90° with respect to the generating voltage. This wave is then propagated towards the collector contact through a drift-free space. When the distance involved is suitably chosen (depending on the velocity of propagation of the charge waves), the output current and voltage will have a phase difference of exactly 180°, so that the diode acts as a negative resistance. With the aid of a microwave resonator, an oscillator can be made. In the p–n version, the breakdown and drift spaces more or less coincide. Continuous output powers of about one watt have been achieved at about 10 GHz in a single device, using diamond heat sinks. As with the Gunn oscillator, combinations of devices are also possible.

A recent development with avalanche oscillators is the so-called Trapatt (Trapped plasma avalanche triggered transit) operation. Very high efficiencies (40–60%) are obtained in pulse operation around a few gigahertz output with powers of about 10 watt. This high efficiency is the result of alternation between high voltage/low current and low voltage/high current conditions in this anomalous mode. With recent, double-drift space, avalanche diodes where both drifting electrons and drifting holes contribute to available output power, a level of about 1 watt at 50 GHz continuous output power has been achieved.

Fig. A3.5 gives the output-power/frequency curves of the various generator types discussed above.

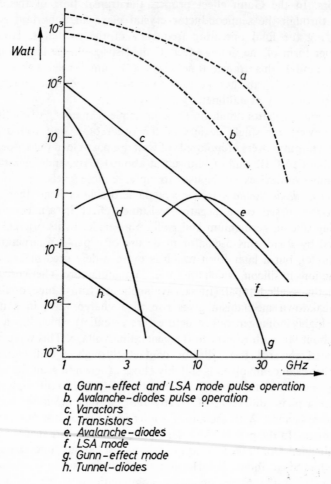

a. Gunn-effect and LSA mode pulse operation
b. Avalanche-diodes pulse operation
c. Varactors
d. Transistors
e. Avalanche-diodes
f. LSA mode
g. Gunn-effect mode
h. Tunnel-diodes

Figure A3.5

Control components

As we have already mentioned in connection with ferrites, a number of control functions, such as attenuation, modulation, switching and limiting, can be carried out with semiconductors; p–i–n diodes and p–n varactors are widely used for these purposes. The impedance of these diodes is a function of the bias: with a 'high' forward voltage the impedance is low, while a negative bias gives a high impedance. In operation, such semiconductor components are placed in a transmission line and subjected to the appropriate voltage pattern, possibly with alternating voltages superimposed on the d.c. voltage.

Semiconductor elements are particularly suitable for low input powers (up to about 100 W continuous), where they have the advantage of fast response. The performance at a given frequency is strongly influenced by parasitic variables such as the housing capacitance, the self-inductance of the lead wires and the series resistance of the semiconductor material itself.

Transmission lines and microminiaturization

The increasing use of semiconductor elements in microwave techniques has had an effect on the types of transmission line used. Very small semiconductor elements can only perform efficiently if the size of the line is changed to suit them. This is partly to avoid parasitic effects, and partly because it is only with small transmission lines that real use can be made of the low weight and small size of semiconductor elements. Sometimes, especially at the lower microwave frequencies, the circuit can again be regarded as a lumped circuit, because the dimensions are once again small compared with the wavelength. Where this is not the case, use is often made of micro-strip lines. These lines are kept small by the use of a substrate such as aluminium oxide, which has a relatively high dielectric constant.

A combination of micro-strip and thin-film techniques, in which a thin conducting film is vacuum-deposited on a suitable substrate to give conducting tracks, makes it possible to use hybrid integrated circuits for microwave components and subsystems, such as mixer stages, the first stage of the intermediate-frequency amplifier and possibly the local oscillator.

Some passive elements (e.g. directional couplers) can be made by a variation of the thin-film technique. This gives a very compact construction, which often has a better performance than the corresponding unit built up of discrete components.

Many of the numerical data in this appendix are taken from laboratory results. Moreover, it should be remembered that the rapid progress being made in this field means that such data are soon out of date.

The wide-scale application in microwave systems of small, cheap solid-state elements with a long life, together with the use of hybrid integrated circuits, can change the whole face of microwave technology, including the range of application of microwaves.

APPENDIX 4   FREQUENCY RANGES AND WAVEGUIDE DIMENSIONS

Frequency ranges

Mainly as a consequence of military applications, a number of microwave frequency bands are often denoted by letters. The limits of these bands are not strictly defined, nor is the use of the letters completely consistent. However, since this form of nomenclature is quite widely used, it is reproduced below, together with the approximate frequency range involved and the characteristic wavelength of each range.

Band designation	Frequency range (GHz)	Characteristic wavelength
P	0·2– 0·4	1 m
L	1 – 2	0·2 m
S	2 – 4	0·1 m
C	4 – 8	50 mm
X	8 –12·5	30 mm
K_u or J	12·5–18	20 mm
K	18 –26·5	12·5 mm
K_a	26·5–40	8 mm
Q	30 –50	8 mm
V	50 –75	4 mm

The various frequency ranges are, moreover, often denoted by descriptive names, again not precisely defined. These are given below, together with the precisely defined CCIR digit code, according to which band n extends from $3 \times 10^{n-1}$ to 3×10^n Hz.

Frequency (Hz)	CCIR code number	Classification
3.10^2		
3.10^3	3	
3.10^4	4	VLF (very low frequency)
3.10^5	5	LF (low frequency)
3.10^6	6	MF (medium frequency)
3.10^7	7	HF (high frequency)
3.10^8	8	VHF (very high frequency)
3.10^9	9	UHF (ultra high frequency)
3.10^{10}	10	SHF (super high frequency)
3.10^{11}	11	EHF (extremely high frequency)

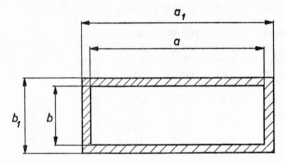

Figure A4.1

Waveguide dimensions

In the table below the dimensions and the various frequency ranges are given of a number of widely used rectangular waveguides.

Wave-guide in IEC nomen-clature	Frequency range in GHz for the principal mode		Internal dimensions in mm			External dimensions in mm		
	From	To	Nominal width a	Nominal height b	Tolerance width and height $\pm$	Nominal width a_1	Nominal height b_1	Tolerance width and height $\pm$
R 3	0·321	0·487	584·2	292·1	—	—	—	—
R 4	0·352	0·535	533·4	266·7	—	—	—	—
R 5	0·41	0·62	457·2	228·60	—	—	—	—
R 6	0·49	0·75	381·0	190·50	—	—	—	—
R 8	0·64	0·98	292·10	146·10	—	—	—	—
R 9	0·76	1·15	247·65	123·82	—	—	—	—
R 12	0·96	1·46	195·58	97·79	—	—	—	—
R 14	1·14	1·73	165·10	82·55	0·33	169·16	86·61	0·20
R 18	1·45	2·20	129·54	64·77	0·26	133·60	68·83	0·20
R 22	1·72	2·61	109·22	54·61	0·22	113·28	58·67	0·20
R 26	2·17	3·30	86·36	43·18	0·17	90·42	47·24	0·17
R 32	2·60	3·95	72·14	34·04	0·14	76·20	38·10	0·14
R 40	3·22	4·90	58·17	29·083	0·12	61·42	32·33	0·12
R 48	3·94	5·99	47·55	22·149	0·095	50·80	25·40	0·095
R 58	4·64	7·05	40·39	20·193	0·081	43·65	23·44	0·081
R 70	5·38	8·18	34·85	15·799	0·070	38·10	19·05	0·070
R 84	6·58	10·0	28·499	12·624	0·057	31·75	15·88	0·057
R 100	8·20	12·5	22·860	10·160	0·046	25·40	12·70	0·05
R 120	9·84	15·0	19·050	9·525	0·038	21·59	12·06	0·05
R 140	11·9	18·0	15·799	7·899	0·031	17·83	9·93	0·05
R 180	14·5	22·0	12·954	6·477	0·026	14·99	8·51	0·05
R 220	17·6	26·7	10·668	4·318	0·021	12·70	6·35	0·05
R 260	21·7	33·0	8·636	4·318	0·02	10·67	6·35	0·05
R 320	26·4	40·1	7·112	3·556	0·02	9·14	5·59	0·05
R 400	33·0	50·1	5·690	2·845	0·02	7·72	4·88	0·05
R 500	39·3	59·7	4·775	2·388	0·02	6·81	4·42	0·05
R 620	49·9	75·8	3·759	1·880	0·02	5·79	3·91	0·05
R 740	60·5	92·0	3·099	1·549	0·02	5·13	3·58	0·05
R 900	73·8	112	2·540	1·270	0·02	4·57	3·30	0·05
R 1200	92·3	140	2·032	1·016	0·02	4·06	3·05	0·05
R 1400	113·6	172·7	1·651	0·826	—	—	—	—
R 1800	144·8	220·0	1·295	0·648	—	—	—	—
R 2200	171·7	261·0	1·092	0·546	—	—	—	—
R 2600	217·1	330·0	0·864	0·432	—	—	—	—

BIBLIOGRAPHY FOR FURTHER READING

Principles of Microwave Circuits, edited by C. G. Montgomery, R. H. Dicke and E. M. Purcell. 1948, McGraw-Hill Book Company, Inc.

Waveguide Handbook, edited by N. Marcuvitz. 1951, McGraw-Hill Book Company, Inc.

Microwave Engineering, by A. F. Harvey. 1963, Academic Press.

Microwave Circuits, by J. L. Altman. 1964, D. Van Nostrand Company, Inc.

Transmission Lines and Waveguides, by L. V. Blake. 1969, John Wiley and Sons, Inc.

Théorie des guides et cavités électromagnétique, by E. Argence and Th. Kahan. 1964, Dunod.

Die Physikalischen Grundlagen der Hochfrequenztechnik, by H. G. Möller. 1955, Springer Verlag.

Microwave Ferrites and Ferrimagnetics, by B. Lax and K. J. Button. 1962, McGraw-Hill Book Company, Inc.

Microwave Filters, Impedance-matching Networks, and Coupling Structures, by G. L. Matthaei, L. Young and E. M. T. Jones. 1964, McGraw-Hill Book Company, Inc.

Microwave Semiconductor Devices and their Circuit Applications, edited by H. A. Watson. 1969, McGraw-Hill Book Company, Inc.

INDEX

Adjustable line length 112
Admittance, on Smith charts 270
Attenuation constant 50
Attenuation measurement 136
Attenuators, waveguide 205

Basic experiments, electrostatics 7–11
  flow of charge 12
Bessel's equation 176
Biot and Savart's law 12
  differential form 12
  integral form 13
Boff diodes 281
Boundary conditions for electromagnetic
    fields 27

CCIR notation 159
Cavity resonators 3, 106
  circular cylinders 236
    as wavemeters 241
  coaxial 248
  coupling of 229
  coupling coefficients 244
  equivalent circuits 244, 248
  introduction 225
  principle of 227
  rectangular parallelepiped 231
  resonance frequency 240
  simple 228
Coaxial lines 67
  frequency range 226
  higher modes in 177
  line constant 88
Complex numbers 45, 253
  basic operations, addition 254
    multiplication 256
  conjugate 255
  logarithms 264
  principal values 259
  series 261
Continuity equations 26

Corkscrew rule 12
Coupling factor 207
Cut-off, in waveguides 156
Cut-off frequency 157

de Moivres' theorem 260
Dielectric constant 10
Dielectric displacement 10
Diodes, avalanche 282
  crystal 127
  Impatt 282
  Read 282
  step recovery 281
  Trapatt 282
Directional isolators 139
Directivity 207
Discontinuities, in coaxial cables 115
Dispersion, in wave transmission 36, 79
Distributed element range 2, 3

Electric current 11
Electric field 49
Electric and magnetic fields, basic
    equations 7
Electric flux 11, 21
Electromagnetic waves 21
  in conducting media 45, 48
  incident on ideal conductors 52
  incident on media with finite
    conductivity 60
  incident on perfect dielectric 56
  polarization 40
    circular and elliptical 41
    plane 40
  propagation of energy 41–3
  in unbounded dielectrics 33
Electrostatics 11
Euler's equation 46

Faraday's law of induction 18–19
Ferrites, in directional isolators 139

291

Field patterns, modes 150
Frequency ranges of waveguides 285

Group velocity, of waveguides 160
Group and phase velocities 79
Guide wavelength 162
Gunn effect oscillators 281

Hertz, Heinrich 22

Impedance, characteristic 76
  of infinitely long transmission lines 82
  input 93
  intrinsic 39
  measurements with transmission lines 107
  in waveguides 166
Impedance transformers, waveguide 188
Interfaces, between dielectrics 31
  between dielectrics and ideal conductors 31–2
  between dielectrics and non-ideal conductors 32
  reflection and transmission at 52

LSA mode 282
Lecher lines 66
  frequency range 226
  line constants 85, 87
  use with valves 226
Line constants 85
  coaxial cables 88
  Lecher line 85, 87
Line stretcher 112
Lumped capacitance in parallel with open line 102
  with short-circuited line 101
Lumped element range 1, 3

Magnetic field 12, 49, 50
Magnetic flux 17, 21, 26
  density (induction) 17
  lines, as closed curves 18
Maxwell's equations 13
  application to interfaces 27
  differential form 22, 25
  integral form 20
  vector form 24, 25
  for waveguides 144
Maxwell's laws, first 14–16, 20, 25
  second 19, 20, 24
Microminiaturization 225
  and transmission lines 284
Mode diagram 240
Mode filters 160

Ohm's law 12
  component notation 25
Oscillators, avalanche diode 282
  Gunn effect 281
  transistor 281

Parallel circuit, formed by open line 103
Parallel wire transmission lines 2
  general properties 82
  propagation in 3
Parasitic capacitances 115
Permeability, relative 17
Phase constant 50
Phase shifter 112
  in waveguides 204
Phase velocity, in waveguides 160, 161
Phase and group velocities 79
Poynting vector 41
Propagation constant 78
  in waveguides 156

Quarter-wave transformer 95, 109
Quartz-crystal oscillators 107

Ratiometers 129
Reflection, in waveguides 162
Reflection coefficient, in transmission lines 89
Refraction, of lines of force 29, 30
Resonance frequency, of cavity resonator 235, 240
Resonators, butterfly 225
  coaxial cavity 248
  coaxial or Lecher line 99, 226
  waveguide 227
Right-hand rule 12

Series circuit formed by open line 103
Short-circuited line, in parallel with open line 102
  series circuit formed by 103
Skin effect 32
Slug tuner 113
Smith chart 109, 265
  for waveguides 173
Solid-state microwave components 275
  (See also Ferrites)
  circulators 276
  control 277, 282
  ferrites 275
  isolators 275
  pre-amplifiers 277
  receivers 277
  semiconductors 277
Standing waves 55
  in transmission lines 88
  in waveguides 162

Standing wave detector 107
  in waveguides 200
Superheterodyne receivers 277
Susceptance 248

T-junctions 209
TE or H waves 146
TM or E waves 151
TE and TM fields, for cavity resonators
  234–5
Tapered sections 113
Taylor series 23
Telegraphers' equations 69, 73
  first and second 74, 75
Transistor amplifiers 279
Transistor oscillators 281
Transmission lines 63
  finite length 88
  impedance measurements with 107
  as impedance transformers 109, 111
  infinite length 82
  parallel wire, attenuation measurement
    136
    attenuators 125–6
    choke plungers 122
    components 119
    detectors 126
    directional couplers 128
    filters 130
    matched loads 123
    power dividers 131
    short-circuit plungers 121
    supports 119
  as resonant circuits 99
  as UHF self-inductances or
    capacitances 96
    open-ended 98
    short-circuited end 96
  as wavemeters 105
Transverse waves 38
Trombone 112
Tunnel-diode amplifiers 278

Varactors 279–81
  punch-through 281
Vector quantities 22, 25
Voltage standing-wave ratio 59, 93

Wave equation 35
  for electric field 50
  general solution 35–6

Wave propagation 21
Waveguides 2, 3
  attenuation in 173
  circular, modes in 176
    propagation in 175
  detection in 199
  dimensions 287
  frequency ranges 285–6
  impedance in 166
  principal modes 159
  propagation in 3
  rectangular 143
    propagation in 143
    propagation control and cut-off
      frequencies 156
  resonators from 227
  slits in 202
  transition to coaxial line 197
Waveguide components, attenuators and
    directional couplers 206
  bends 194
  capacitive rod 187
  capacitive screen 183
  discontinuities 179
  filters 215
    band-pass 219
    band-stop 218
    high-pass 215
  flanges 221
  impedance transformers 188
  inductive iris 182
  inductive rod 186
  inductive screen, finite thickness 186
  matched loads 205
  phase shifters 204
  power dividers 213
  quarter-wave impedance transformers
    191
  resonant aperture 184
  rod of variable length 187
  screen with small round hole 184
  short-circuit plungers 194
  sliding screw 189
  slug tuners 190
  standing wave detectors 200
  stub tuners 190
  T-junctions 209
  tuning screw 189
Waveguide connections 3
Wavelength, and size of circuit 1
Wavemeters 105
  cylindrical cavity resonators as 241